Knowledge System
and Curriculum
Setting for
Undergraduate
Program of
Energy Storage
Science and
Engineering

储能科学与
工程本科专业
知识体系与课程设置

● 西安交通大学储能科学与工程专业建设组
● 主编 何雅玲

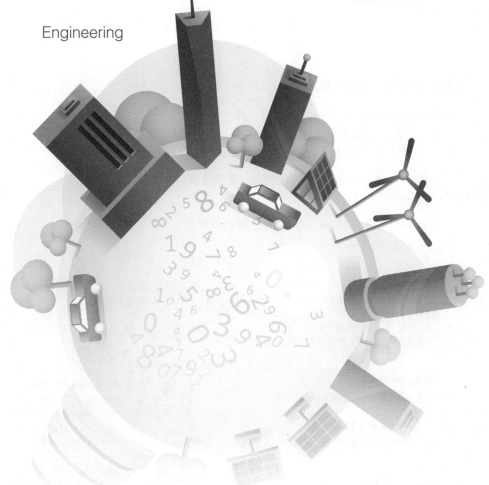

高等教育出版社·北京

内容简介

本书针对高等学校储能科学与工程本科专业人才培养的专业内涵、定位和知识体系,设置了数学基础课程群、储能基础课程群、热质储能课程群、电化学与电磁储能课程群、系统储能课程群、前沿讲座课程群、专业综合实验课程群,重点介绍了这7大课程群中各门课程的概况和知识点,为培养具有多学科交叉素养的科研工作者、工程师及管理人员奠定知识和能力的基础。

本书可为高等学校储能科学与工程专业本科阶段的课程体系提供指导,同时也可为能源动力类、电气类、电子信息类、机械类、物理学类、化学类、力学类、材料类、自动化类、计算机类、水利类、化工与制药类、交通运输类、航空航天类、核工程类、环境科学与工程类、建筑类等专业构建宽口径和学科交叉的课程体系提供参考和引导示范,还可为上述专业研究生相关课程体系的建设和专业学习提供参考。

图书在版编目(CIP)数据

储能科学与工程本科专业知识体系与课程设置 / 何雅玲主编. --北京:高等教育出版社,2021.7
ISBN 978-7-04-056142-5

Ⅰ. ①储… Ⅱ. ①何… Ⅲ. ①储能-课程设置-高等学校 Ⅳ. ①TK02

中国版本图书馆CIP数据核字(2021)第098606号

Chuneng Kexue Yu Gongcheng Benke Zhuanye Zhishi Tixi Yu Kecheng Shezhi

策划编辑	宋 晓	责任编辑	宋 晓	封面设计	姜 磊	版式设计	徐艳妮
责任校对	马鑫蕊	责任印制	田 甜				

出版发行	高等教育出版社	网　　址	http://www.hep.edu.cn	
社　　址	北京市西城区德外大街 4 号		http://www.hep.com.cn	
邮政编码	100120	网上订购	http://www.hepmall.com.cn	
印　　刷	北京鑫海金澳胶印有限公司		http://www.hepmall.com	
开　　本	787 mm×1092 mm　1/16		http://www.hepmall.cn	
印　　张	24			
字　　数	560 千字	版　　次	2021 年 7 月第 1 版	
购书热线	010-58581118	印　　次	2021 年 7 月第 1 次印刷	
咨询电话	400-810-0598	定　　价	51.00 元	

本书如有缺页、倒页、脱页等质量问题,请到所购图书销售部门联系调换
版权所有　侵权必究
物 料 号　56142-00

又踏层峰望眼开

　　能源安全是关系国家经济社会发展的全局性、战略性问题，能源战略的成败，关乎国家民族命运。习近平总书记从国家发展和安全的战略高度，对推动能源消费、能源供给、能源技术和能源体制革命做出了一系列重要部署。储能技术作为新能源发展的核心支撑，在促进能源生产消费、开放共享、灵活交易、协同发展，推动能源革命和能源新业态发展等方面发挥着至关重要的作用。创新突破的储能技术将成为带动全球能源格局革命性、颠覆性调整的重要引领技术，世界主要发达国家纷纷加快发展储能产业，大力规划建设储能项目，加强储能产业人才培养和技术储备，抢占能源战略突破的制高点。因此，加快建立与发展储能科学与工程专业及学科，加快培养急需紧缺人才，破解储能技术的共性和瓶颈技术难题，是推动我国储能产业和能源产业高质量发展的迫切需要和战略选择。

　　在教育部、国家发展和改革委员会、国家能源局等的支持下，西安交通大学经过一年多的细致筹备，于2020年2月设立了我国第一个且当年唯一一个"储能科学与工程专业"。在储能科学与工程专业筹建过程中，西安交通大学各级领导给予了大力支持，教务处进行了精心筹划、协调、组织，相关学院教师都做了大量细致深入的工作。从专业建设的内涵讨论、培养体系的建设、师资力量的配备等各个环节，到每一门课程的定制，甚至某一个知识点取舍的讨论，始终都围绕一个问题：如何才能做到真正的学科交叉，充分发挥储能技术的优势？　一次又一次的思考和研讨，一遍又一遍的成案与修改，让我们深感储能科学与工程专业学科交叉的重要性及魅力所在。

　　储能科学与工程专业是一个多学科深度交叉融合的新兴专业，它具有多样化的应用背景、多维度的知识体系、多层次的专业构架等特征，涉及化学、物理学、动力工程及工程热物理、材料科学与工程、电气工程、化学工程与技术、机械工程等众多学科。在多学科交叉融合的基础上，目前形成了数学基础课程群、储能基础课程群、热质储能课程群、电化学与电磁储能课程群、系统储能课程群、前沿讲座课程群与专业综合实验课程群7大课程群，后续我们将努力加强完善产教融合支撑体系，进一步优化课程群体系，使之更完备、更科学、更加适应能源战略的需要。

　　本书付梓之时，正值新冠肺炎肆虐全球，国际形势波谲云诡，世界正经历百年未有之大变局。此时此景，再思考储能科学与工程专业人才培养之内涵，更感责任重大、意义深远。习近平总书记强调："发展是第一要务，人才是第一资源，创新是第一动力。"千秋基业，人才为先。通过对储能科学与工程专业需要掌握的基础知识、理论、方法进行科学的规划设计，汇聚学术声望高、专业理论水平扎实、实践教学经验丰富的精英师资队伍，努力培养储能领域具有全球化视野的科技、工程、技术管理复合型的高端人才。

　　从实践中来，到实践中去。储能科学与工程这个新兴专业顺应能源革命的实践而生，也将在实践中不断丰富和发展，相应的人才培养模式也应该是开放式的、启发式的，需要在实践中不断发展和完善。科学思维的培养，创新火花的点燃，尤胜于具体知识点的传授。"知之者不如好之者，好之者不如乐之者"，真正让学生理解这个专业的重要性，发自内心地热爱这个专业，才是我们的初心和使命。

在西安交通大学多个一流学科强有力支撑的基础上，在一大批优秀教师的倾情付出中，储能科学与工程专业就像站在巨人肩上的新秀，璀璨而生，起点高，基础好，发展快。相信她会发展成为一个在国内外具有影响力、能为我国培养大批优秀人才的金牌专业！

又踏层峰望眼开，且待佳音破晓来。我们对此充满期盼，并为之奋斗！

西安交通大学

2020 年 10 月 1 日于西安

西安交通大学储能科学与工程专业建设专家委员会

顾　　问

陶文铨（中国科学院院士，西安交通大学能源与动力工程学院）

首席科学家

何雅玲（中国科学院院士，西安交通大学能源与动力工程学院）

委　　员

管晓宏（中国科学院院士，西安交通大学自动化科学与工程学院）

别朝红（西安交通大学电气工程学院）

徐友龙（西安交通大学电子科学与工程学院）

严俊杰（西安交通大学能源与动力工程学院）

丁书江（西安交通大学化学学院）

唐桂华（西安交通大学能源与动力工程学院）

李印实（西安交通大学能源与动力工程学院）

宋政湘（西安交通大学电气工程学院）

杨生春（西安交通大学物理学院）

宋江选（西安交通大学材料科学与工程学院）

兰　剑（西安交通大学自动化科学与工程学院）

项 目 主 任

兰　剑（西安交通大学自动化科学与工程学院）

西安交通大学储能科学与工程本科专业知识体系建设与课程设置
工 作 组

郑庆华（西安交通大学副校长）

何雅玲（西安交通大学能源与动力工程学院，中国科学院院士）

徐忠锋（西安交通大学教师教学发展中心主任）

兰　剑（西安交通大学自动化科学与工程学院，教务处副处长）

别朝红（西安交通大学电气工程学院）

徐友龙（西安交通大学电子科学与工程学院）

宋政湘（西安交通大学电气工程学院）

丁书江（西安交通大学化学学院）

宋江选（西安交通大学材料科学与工程学院）

唐桂华（西安交通大学能源与动力工程学院）

李印实（西安交通大学能源与动力工程学院）

席　奂（西安交通大学能源与动力工程学院）
陈雪江（西安交通大学能源与动力工程学院）
杨生春（西安交通大学物理学院）
杨清宇（西安交通大学自动化科学与工程学院）
唐亚哲（西安交通大学计算机科学与技术学院）
唐　伟（西安交通大学化学工程与技术学院）
吴　江（西安交通大学自动化科学与工程学院）

西安交通大学储能科学与工程本科专业教学大纲制定成员

主　编

何雅玲

编写组成员

数学基础课程群

数学与统计学院：李换琴，高安喜，刘峰，赵小艳，吴慧卓

自动化科学与工程学院：翟桥柱，徐占伯

储能基础课程群

化学学院：丁书江，张雯，李银环，徐四龙，白艳红

物理学院：杨生春，王小力，李宏荣，赵铭姝

电气工程学院：刘文凤，罗先觉，宋政湘，赵进全

能源与动力工程学院：唐桂华，陶于兵

航天航空学院：吴莹，解社娟，文毅

计算机科学与技术学院：张兴军

自动化科学与工程学院：管晓宏，张爱民，杨清宇

机械工程学院：段玉岗，续丹，许睦旬

热质储能课程群

能源与动力工程学院：陶文铨，何雅玲，王秋旺，唐桂华，李印实，李明佳，陶于兵，
秦国良，陈黎，席奂，唐上朝

电化学与电磁储能课程群

材料科学与工程学院：宋江选，宁晓辉

电子科学与工程学院：徐友龙

化学学院：丁书江，高国新

电气工程学院：韩晓刚，王鹏飞，沈飞

系统储能课程群

电气工程学院：王锡凡，别朝红，王建学，杨旭，宋政湘

自动化科学与工程学院：兰剑

计算机科学与技术学院：张兴军

自动化科学与工程学院：管晓宏，徐占伯

前沿讲座课程群

能源与动力工程学院：陶文铨，何雅玲，唐桂华，李印实

电子科学与工程学院：徐友龙

电气工程学院：宋政湘

专业综合实验课程群

能源与动力工程学院：李明佳，席奂

电子科学与工程学院：徐友龙

电气工程学院：宋政湘，孙丽琼

英文校审

钱苏昕，李明佳

全书统稿

何雅玲，席奂

前　言

随着全球能源格局由传统化石能源向清洁高效能源的深刻转变，我国能源结构也正经历着前所未有的深层次变革，新能源在能源供给侧占有越来越重要的战略地位。储能技术和储能产业作为新能源发展的核心支撑，覆盖电源侧、电网侧、用户侧、居民侧以及社会化功能性储能设施等多方面需求，贯穿新能源开发与利用的全部环节，储能设施的加快建设将成为我国构建更加清洁低碳、安全高效的现代能源产业体系的重要基础。因此，储能技术既是国家能源安全的重要保障，也是电动汽车等新兴产业的主要发展动力，具有重要的战略意义和不可估量的产业前景。

2014 年 6 月 13 日，习近平总书记在中央财经领导小组第六次会议上明确提出 "四个革命、一个合作" 的国家能源安全战略思想。国家发展和改革委员会与国家能源局于 2016 年联合发布了《能源技术革命创新行动计划（2016—2030 年）》和《能源生产和消费革命战略（2016—2030）》，明确指出加强先进储能技术创新，促进能源与现代信息技术深度融合，全面建设 "互联网＋" 智慧能源，推动互联网与分布式能源技术、先进电网技术与储能技术深度融合，抢占能源技术进步先机，抢占新一轮科技革命和产业变革竞争制高点。为实现这一目标，国家加快培养储能科学与技术方面的高端人才，鼓励高校开展储能科学与工程专业教育，加强产学研合作，鼓励高校、科研院所与企业等机构合作开展先进储能学科建设。因此，创办储能科学与工程专业，是贯彻国家《能源技术革命创新行动计划（2016—2030 年）》和《能源生产和消费革命战略（2016—2030）》的重要举措。

目前尚未见到国内外大学将储能技术设立为单独专业的报道，仅在能源相关各专业中进行方向性的人才培养，难以支撑储能技术迅猛发展对人才的迫切需求。为加快培养储能领域 "高、精、尖、缺" 人才，增强产业关键核心技术攻关和自主创新能力，以产教融合发展推动储能产业高质量发展，教育部、国家发展和改革委员会、国家能源局联合制定了《储能技术专业学科发展行动计划（2020—2024 年）》；2020 年 2 月，教育部新增设了 "能源动力类—储能科学与工程" 特设专业，并批准西安交通大学开设全国首个且当年唯一一个储能科学与工程专业。2019 年，在学校领导的大力支持下，由中国科学院院士何雅玲教授领衔，西安交通大学创办了储能科学与工程本科专业，并于 2020 年从在校的众多报名大学生中，经过严格选拔，招收了第一批储能科学与工程专业的本科生。

储能科学与工程专业是一个多学科高度交叉融合的新专业，人才培养具有一定的挑战性，其课程体系必须充分考虑以学科交叉为指导思想。自 2019 年开始，我们经过近 2 年的深入研究和反复研讨，结合我国新工科建设和国际工程教育发展趋势，最终形成了这本书中给出的储能科学与工程本科专业的知识体系和课程设置，课程包括数学基础课程群、储能基础课程群、热质储能课程群、电化学与电磁储能课程群、系统储能课程群、前沿讲座课程群及专业综合实验课程群 7 大课程群，共计 43 门课程，其中必修课 31 门，选修课 12 门；在实践方面还专门设置了专业综合实验课程群。经过遴选，从事储能科学与工程专业相关教学工作的教师，分布于西安交通大学能源与动力工程学院、电气工程学院、电子科学与工程学院、物理学院、化学学院、计算机科学与技术学院、自动化科学与工程学院、材料科学与工程学院、机械工程学院、航天航空学院、数学与统计学院等多个学院，并聘请国内知名高校、科研机构和储能行业知名专家加入。如此雄厚的师资力量为储能科学与工程专业的本科生培养提供了坚实的师资保障。

为了办好储能科学与工程专业，我们编著了本书，通过课程体系对储能科学与工程专业本科人才培养工作给予引导，培养学生拥有扎实宽广的储能基础理论与专业知识以及多学科综合知识，具有整合思维、工程推理和解决复杂工程问题的能力，具备从事储能材料、器件与系统的研究、开发、设计、制造和管理的技术能力，具有敏锐的创新意识与出色的创新能力，能够适应未来各储能产业领域对高素质复合型创新人才的需求。

高等教育出版社为本书的出版剪莽拥篲，给予了大力支持，在此深表感谢！

作为国内第一个储能科学与工程专业，其知识体系与课程设置没有前事可鉴，虽然编者殚精竭虑，本书难免存在疏漏和不足，殷切希望广大读者批评指正，以利于我们不断地改进完善，这是对我们工作的最大帮助和支持，衷心感谢！

西安交通大学

何雅玲　陶文铨　郑庆华　管晓宏

2020 年 12 月

目 录

储能科学与工程专业及人才培养定位

1.1　储能科学与工程

专业代码：080504T

授予学位：工学学士

学制：4 年

储能科学与工程专业以能量的存储即储能技术为研究对象。储能技术在促进能源生产消费、开放共享、灵活交易、协同发展，推动能源革命和能源新业态发展方面发挥着至关重要的作用；也是能源转换与缓冲、调峰与提效、传输与调度、管理与运用等环节中的核心技术。储能技术的创新突破将成为带动全球能源格局革命性、颠覆性调整的重要因素。

本专业所研究的储能包括机械储能、热质储能、电化学与电磁储能、系统储能等。对各种储能技术的原理、储能材料的开发与研制、储能设备的研制与制造、储能系统的设计及运行策略、其经济性和大数据分析等知识，进行系统的学习和研究。储能科学与工程是一个多学科深度交叉融合的新兴专业，它具有多样化的应用背景、多维度的知识体系、多层次的专业构架等特征，涉及化学、物理学、能源与动力工程、材料科学与工程、电气工程及其自动化、化学工程与工艺、机械工程等众多学科的深度交叉和融合。

1.2　人才培养需求

近年来，我国高度重视储能科学与技术的发展。2014 年习近平总书记提出了新时代能源战略思想，2016 年国家发展和改革委员会与国家能源局联合发布了《能源技术革命创新行动计划（2016—2030 年）》和《能源生产和消费革命战略（2016—2030）》，2017年国家发展和改革委员会、财政部、科学技术部、工业和信息化部及国家能源局联合颁布了《关于促进储能技术与产业发展的指导意见》，2020 年教育部、国家发展和改革委员会与国家能源局三部委联合颁布了《储能技术专业学科发展行动计划（2020—2024 年）》，教育部 2020 年 2 月增设了"能源动力类——储能科学与工程"特设专业。这些充分证明了储能技术在我国能源革命中占有非常重要的地位。

2020 年 9 月 22 日，习近平总书记在第 75 届联合国大会一般性辩论上宣布我国力争于 2030 年前二氧化碳排放达到峰值的目标以及努力争取于 2060 年前实现碳中和的愿景，并在气候雄心峰会上进一步宣布国家自主贡献最新举措。这在全世界面前体现了中国负责任大国的国际形象，同时对我国能源革命提出了更严峻的任务和目标。

在这迫切的形势下，加快推进储能技术和学科的建设，培养储能方面"高、精、尖、缺"人才，对保障我国能源革命的顺利进行，具有重大的战略意义。

1.3　人才培养定位

西安交通大学储能科学与工程专业面向国家能源战略重大需求，顺应我国新工科建设和国际工程教育发展新趋势，坚持以学生为中心、产出为导向的工程教育理念，以"数理基础厚实、专业交叉融合、工程思维导向、实践能力创新、个性模块管理"为特色，依托和整合西安交通大学 6 个理工类优势学科（其中两个学科排名全国第一），针对储能系统设计、应用及安全运维，高性能储能材料设计与制备，技术经济性和大数据分析等领域，培养站在世界储能技术前沿、勇于创新的技术领军人才与具有宏观战略思维和市场思维的复合型管理人才；本专业同时支撑储能科学与工程学科建设，实施产教融合，推动提高我国储能行业的自主创新和关键核心技术攻关能力，实现储能产业健康快速发展。

储能科学与工程专业培养方案

2.1　培养目标

　　面向国家能源结构改革及储能产业发展的重大需求，致力于培养德、智、体、美、劳全面发展，掌握储能专业基础理论知识和专业技能方面的多学科综合知识，具有科学素养、实践能力、创新能力、系统思维能力、工程推理和解决复杂工程问题能力、产业视角与国际视野，具备从事储能材料、器件与系统的研究、开发、设计、制造和管理的技术能力和工程实践能力，有潜力成长为国际一流工程师、科学家及企业家，能在我国储能科技、储能产业技术发展中发挥优秀领军作用的高素质、复合型人才。

2.2　培养方式

　　西安交通大学储能科学与工程专业实施由 6 大学科交叉（能源动力、电气、电子信息、化学、物理学、材料）、多个龙头企业协同、国际前沿引领的"通专融合"拔尖创新人才培养新模式。本科生选拔遵循"优中选优"的原则，将兴趣、能力与潜力作为选拔与评价的依据，遴选出数学、物理学、化学等学科基础扎实、能力突出、对储能科学与工程兴趣浓厚、具有良好发展潜质的优秀学生进入储能科学与工程专业学习。

　　课程体系的建设在参考借鉴国际一流大学、机构近似专业课程设置与培养理念的同时，充分把握储能科学与工程学科仍处于高速发展期、渗透性与学科交叉性强的特点，课程设置精练，选用国际一流教材和自编高水平教材，通过讲授基本知识锻炼学习能力与思维方法，并让学生拥有自主学习和创造知识的空间。

　　任课教师的聘用坚持结构合理、专兼结合的"双师型"师资队伍建设原则。师资吸引来自能源与动力工程学院、电气工程学院、数学与统计学院、物理学院、化学学院、电子与信息学部、材料科学与工程学院、化学工程与技术学院、机械工程学院、前沿科学技术研究院等多个学院/部的一批优秀教师队伍，集优势学科，聚前沿发展，重交叉创新。师资队伍中包括院士、国家级教学名师、长江学者、国家杰出青年在内的一批教师，此外还包括来自国内著名企业的行业专家。雄厚的师资力量为储能科学与工程专业的本科生培养提供了坚实的师资保障。

　　不断改革、创新教学方法。采用案例式教学、启发式教学等模式，加强与学生的讨论和互动；改革考试、考核方式，重点激发学生的想象力和批判性思维，培养学生的表达能力、发现问题的能力和学术判断力；结合学科最新研究成果，充分发挥科研与教学相互促进的作用,通过导师个性化指导和科研训练项目,引导学生发现学术兴趣、选择科研方向、发展学术特长。

加强实验环节的建设和改革。通过课内实验、专业综合实验、项目训练和导师制科研训练、工业实习、国际交流等实践环节,培养学生的学习兴趣,提升学习的积极性及灵活运用所学知识解决实际问题的能力,激发创新思维和能力。

产教深度融合,校企协同育人。加强与一流企业的合作,从育人理念、合作机制、组织方式、实现路径上不断探索与创新,构建"产教融合、校企合作、协同育人"新生态,培养高素质和具有产业应用视角的创新性人才。

加强国际化培养力度。开展国际合作,建立多层次、立体化的高端国际联合培养体系,通过与国际一流大学和研究机构合作办学、资助优秀学生国(境)外交流学习或参加高水平国际会议、邀请国际一流学者访问讲学等,形成国际化培养氛围,拓宽学生的国际化视野。

2.3　专业知识体系

储能科学与工程本科专业的知识体系和课程设置主要由数学基础课程群、储能基础课程群、热质储能课程群、电化学与电磁储能课程群、系统储能课程群、前沿讲座课程群及专业综合实验课程群 7 大课程群构成。

数学基础课程群中除了有关于数学和统计的通识课程外,还设置了"数学物理方程"和"运筹学"等课程。储能基础课程群中除了大学物理、工程力学、计算机科学基础与高级程序设计、电路、工程图学以外,还设置了"现代电子技术""自动控制理论""储能化学基础""储能原理"等课程。数学基础与储能基础两个课程群作为储能科学与工程专业的基础课程,可以为学生打下坚实而宽广的学科基础。

另外,针对储能领域中热质储能模块、电化学与电磁储能模块、系统储能模块 3 个不同模块方向的特点,分别设置了独立的课程群。热质储能课程群由"储能热流基础""传热传质学""热质储能技术及应用""可再生能源及其发电技术""氢能储存与应用""流体机械原理及其储能应用"等课程组成,以启发学生探索热质储能的未来与奥秘。电化学与电磁储能课程群由"储能材料工程""电化学基础""纳米材料与能源""半导体物理""新型储能电池技术"和"固态电池"等课程组成,介绍了通过电化学反应的方式实现能量储存的技术。系统储能课程群由"储能系统设计""电力系统分析""能源互联网""嵌入式智能系统""智能电网储能应用技术""电储能系统与并网技术""信息物理融合能源系统"等课程组成,以培养学生应用储能技术的能力。这些课程群设置了选修课与必修课,围绕 3 个储能模块,除了为学生奠定具有不同特色的专业核心知识根基外,也充分考虑了学生知识面的拓展,希望学生跨模块选择课程。

前沿讲座课程群为学生提供前沿的储能科学与技术知识,介绍最新的研究动态和开发成果,使学生具备良好的学科前沿视野,掌握行业最新动态。专业综合实验课程群包含"储能电池设计、制作及集成化实验""储能装置设计与开发实验"等课程,通过案例式教学、实际项目完整的开发实现过程,培养学生解决实际问题的能力,锻炼学生组织、管理、写作、表达和动手能力等。

2.4 专业课程设置

专业课程设置了数学基础课程群、储能基础课程群、热质储能课程群、电化学与电磁储能课程群、系统储能课程群、前沿讲座课程群及专业综合实验课程群等 7 大课程群，共计 43 门课程，其中必修 31 门，选修 12 门。其中专门设置了专业综合实验课程群，用以培养学生综合运用所学知识动手解决实际问题的能力，提高学生的创新素质。7 大课程群内所有课程名称、学分情况及课程设置情况等见表 2.1。

表2.1 7大课程群内课程名称等情况介绍

课程群	课程名称（学分）	必/选修	参阅章节
数学基础课程群 （必修28学分）	高等数学Ⅰ（13）	必修	3.1
	线性代数与解析几何（4）		3.2
	概率论与数理统计（3）		3.3
	复变函数与积分变换（3）		3.4
	数学物理方程（2）		3.5
	运筹学（3）		3.6
储能基础课程群 （必修44.5学分）	大学物理（含实验）（10）	必修	4.1
	工程力学（3）		4.2
	储能化学基础（含实验）（7.5）		4.3
	计算机科学基础与高级程序设计（6）		4.4
	自动控制理论（3）		4.5
	工程图学（3）		4.6
	电路（4）		4.7
	现代电子技术（3）		4.8
	储能原理Ⅰ（3）		4.9
	储能原理Ⅱ（2）		4.10
热质储能课程群 （必修10.5学分、 选修8学分）	储能热流基础（含实验）（4.5）	必修	5.1
	传热传质学（3）		5.2
	热质储能技术及应用（含实验）（3）		5.3
	先进热力系统技术及仿真（2）	选修	5.4
	可再生能源及其发电技术（2）		5.5
	流体机械原理及其储能应用（2）		5.6
	氢能储存与应用（2）		5.7
电化学与电磁储能课程群 （必修9学分、选修8学分）	储能材料工程（3）	必修	6.1
	电化学基础（3）		6.2
	半导体物理（3）		6.3

续表

课程群	课程名称（学分）	必/选修	参阅章节
电化学与电磁储能课程群 （必修9学分、选修8学分）	电池材料制备技术（2）	选修	6.4
	新型储能电池技术（2）		6.5
	纳米材料与能源（2）		6.6
	固态电池（2）		6.7
系统储能课程群 （必修13学分、 选修8学分）	储能系统设计（2）	必修	7.1
	电力系统分析（5）		7.2
	储能系统检测与估计（2）		7.3
	能源互联网（2）		7.4
	嵌入式智能系统（2）		7.5
	智能电网储能应用技术（2）	选修	7.6
	电储能系统与并网技术（2）		7.7
	物联网应用概论（2）		7.8
	信息物理融合能源系统（2）		7.9
前沿讲座课程群 （必修1学分）	大型储能工程导论（1）	必修	8.1
专业综合实验课程群 （必修3学分）	热质储能综合实验（1）	必修	9.1
	储能电池设计、制作及集成化实验（1）		9.2
	储能装置设计与开发实验（1）		9.3

2.5　学期安排

　　学期安排建议见表 2.2。需要特别说明的是，表 2.2 中不含普通高等学校统一要求的通识教育、集中实践等公共课程，此类课程可根据学校具体要求安排至相应学期。

表2.2　学期安排建议

	第一学期：1-1		第二学期：1-2		小学期1
	课程名称（学分）	参阅章节	课程名称（学分）	参阅章节	
第一学年	高等数学Ⅰ（第一部分）(6.5)	3.1	高等数学Ⅰ（第二部分）(6.5)	3.1	前沿讲座
	线性代数与解析几何（4）	3.2	大学物理（含实验） （第一部分）(5)	4.1	
	储能化学基础（第一部分） （含实验）(4)	4.3	储能化学基础（第二部分）（含实验）(3.5)	4.3	

续表

	第三学期：2-1		第四学期：2-2		小学期2
	课程名称（学分）	参阅章节	课程名称（学分）	参阅章节	
第二学年	概率论与数理统计（3）	3.3	计算机科学基础与高级程序设计（第一部分）（2）	4.4	前沿讲座
	复变函数与积分变换（3）	3.4	工程图学（3）	4.6	
	数学物理方程（2）	3.5	现代电子技术（3）	4.8	
	大学物理（含实验）（第二部分）（5）	4.1	储能原理Ⅰ（3）	4.9	
	工程力学（3）	4.2	储能热流基础（含实验）（4.5）	5.1	
	电路（4）	4.7	电化学基础（3）	6.2	
	第五学期：3-1		第六学期：3-2		小学期3
	课程名称（学分）	参阅章节	课程名称（学分）	参阅章节	
第三学年	运筹学（3）	3.6	热质储能技术及应用（含实验）（3）	5.3	前沿讲座
	计算机科学基础与高级程序设计（第二部分）（4）	4.4	传热传质学（3）	5.2	
	自动控制理论（3）	4.5	储能系统检测与估计（2）	7.3	
	储能原理Ⅱ（2）	4.10	能源互联网（2）	7.4	
	储能材料工程（3）	6.1	嵌入式智能系统（2）	7.5	
	半导体物理（3）	6.3	智能电网储能应用技术（2）	7.6	
	储能系统设计（2）	7.1	电储能系统与并网技术（2）	7.7	
	电力系统分析（5）	7.2	信息物理融合能源系统（2）	7.9	
	第七学期：4-1		第八学期：4-2		小学期4
	课程名称（学分）	参阅章节			
第四学年	热质储能综合实验（1）（专业综合实验三选一）	9.1	毕业设计等		
	储能电池设计、制作及集成化实验（1）（专业综合实验三选一）	9.2			
	储能装置设计与开发实验（1）（专业综合实验三选一）	9.3			
	先进热力系统技术及仿真（2）	5.4			
	可再生能源及其发电技术（2）	5.5			

第四学年	第七学期：4-1		第八学期:4-2	小学期4
	课程名称（学分）	参阅章节		
	流体机械原理及其储能应用（2）	5.6	毕业设计等	
	氢能储存与应用（2）	5.7		
	电池材料制备技术（2）	6.4		
	新型储能电池技术（2）	6.5		
	纳米材料与能源（2）	6.6		
	固态电池（2）	6.7		
	物联网应用概论（2）	7.8		

2.6　毕业要求

西安交通大学储能科学与工程专业 7 大课程群共需修 109 学分，其他通识教育、集中实践等公共课程需修 56 学分。以上合计需修 165 学分，达到储能科学与工程专业的毕业要求。

储能科学与工程专业学制为 4 年，毕业将授予工学学士学位。

数学基础课程群

3.1 "高等数学Ⅰ"教学大纲

课程名称：高等数学Ⅰ
Course Title：Advanced Mathematics Ⅰ
先修课程：无
Prerequisites：None
学分：13
Credits：13

3.1.1 课程目的和基本内容（Course Objectives and Basic Contents）

本课程是储能科学与工程专业本科生的基础必修课。

本课程介绍了极限、微分、积分、级数等重要的数学工具，并将分析、代数和几何内容进行了有机结合。相关的知识对包括储能科学与工程专业在内的众多工科专业提供了不可或缺的高等数学基础，也使学生在数学的抽象性、逻辑性和严谨性等方面受到必要的熏陶和训练。可为学生今后增进数学知识、学习相关专业知识等奠定良好的基础，同时也可培养学生应用数学知识进行数据分析、建模及解决实际问题的能力。

教学目的：

（1）要求学生系统地掌握一元函数微积分学、无穷级数、多元函数微积分学、常微分方程组的基本概念、基本理论和基本方法等，同时通过数学实验来培养学生的综合素质，即实验动手能力、分析设计能力及团队合作精神，拓展学生思维，激发学生的创新意识。

（2）结合数学实验模块，训练学生的建模意识、能力以及使用数学软件求解所建立模型的能力，学会对实际问题进行建模，进而进行模拟和分析，最终解决问题。

（3）提升学生的数学素质，使学生了解数学的前沿发展，特别是在工程技术方面的最新应用。

This course is a fundamental compulsory course for undergraduates in energy storage science and engineering.

This course introduces some important mathematical tools, such as limit, differential, integral and series, and organically combines analysis, algebra and geometry. Relevant knowledge in this course provides an indispensable foundation of advanced mathematics for energy storage science and engineering majors and lots of other engineering majors. It also lets students take necessary trainings in abstraction, logic and rigor of mathematics. In doing so, it not only enables students to lay a good foundation to pursue more mathematics knowledge and relevant professional knowledge, but also cultivates students' ability to apply mathematics knowledge to perform data analysis model and solve practical problems.

Instructional objectives:

（1）This course requires students to systematically master basic concepts, basic theories and basic methods of single variable calculus, infinite series, multivariable calculus , and ordinary differential equations, etc. Meanwhile, the comprehensive quality, which includes experimental hands-on ability, analysis design ability and teamwork spirit, of students can be cultivated through mathematical experiments. In doing so, students' critical thinking are enhanced or expanded and their sense of innovation are inspired.

（2）Combining with the mathematical experiment module, this course trains students' modeling awareness, and the ability to solve the built models by using mathematical software. Students should learn how to model practical problems , and then do the simulations and analysis , and finally solve the problems.

（3）Improve the students' mathematical quality and make them learn the frontier developments of mathematics, especially their latest applications in engineering technology.

3.1.2　课程基本情况（Basic Information of the Course）

课程名称	高等数学 I Advanced Mathematics I											
开课时间	一年级		二年级		三年级		四年级		总学分	13	总学时	220
	秋	春	秋	春	秋	春	秋	春				
课程定位	储能科学与工程专业本科生数学基础课程群必修课											
授课学时分配	课堂讲授196学时+实验24学时											
先修课程	无											
后续课程	概率论与数理统计,线性代数与解析几何,数学物理方程,复变函数与积分变换,运筹学,大学物理（含实验）,工程力学,电路,现代电子技术,储能原理 I ,储能原理 II ,储能热流基础（含实验）,传热传质学,流体机械原理及其储能应用,电力系统分析,储能系统检测与估计,物联网应用概论											
教学方式	课堂教学,讨论,作业											
考核方式	课程结束笔试成绩占60%,实验报告占10%,平时成绩占10%,期中考试占20%											

参考教材	[1] 王绵森,马知恩.工科数学分析基础. 3版.北京:高等教育出版社,2017. [2] 李继成.数学实验. 2版.北京:高等教育出版社,2014.
参考资料	[1]武忠祥.工科数学分析基础教学辅导书.北京:高等教育出版社,2006.

3.1.3 教学目的和基本要求(Teaching Objectives and Basic Requirements)

(1)培养学生的创新能力,提升学生的综合素质。

(2)培养学生的数学抽象思维能力、严密的逻辑推理能力和丰富敏感的空间想象能力。

(1) Cultivate students' innovation ability and improve their comprehensive quality.

(2) Cultivate students' ability of abstract thinking in mathematics, the ability of logical reasoning, and the ability of spatial imagination.

3.1.4 课程大纲和知识点(Syllabus and Key Points)

第一章 函数、极限、连续(function, limit and continuity)

章节序号 chapter number	章节名称 chapters	知识点 key points	课时 class hour
1.1	集合、映射与函数 sets, mappings and functions	(1)集合及其运算 (2)实数集的完备性与确界存在定理 (3)映射与函数的概念 (4)线性函数的基本属性 (5)复合映射与复合函数 (6)逆映射与反函数 (7)初等函数与双曲函数 (1) sets and their operations (2) completeness of the real numbers and the theorem on existence of supremum and infimum (3) concepts of mappings and functions (4) basic characteristics of linear functions (5) composition of mappings and composition of functions (6) inverse mappings and inverse functions (7) elementary functions and hyperbolic functions	3

续表

章节序号 chapter number	章节名称 chapters	知识点 key points	课时 class hour
1.2	数列极限 limit of sequence	（1）数列极限的概念 （2）收敛数列的性质 （3）数列收敛的判别准则 （1）concept of limit of a sequence （2）properties of convergent sequences （3）conditions for convergence of a sequence	6
1.3	函数极限 limit of function	（1）函数极限的概念 （2）函数极限的性质 （3）两个重要极限 （4）函数极限的存在准则 （1）concept of limit of a function （2）properties of function limits （3）two important limits （4）existence criteria for the limit of a function	3
1.4	无穷小量与无穷大量 infinitely small quantities and infinitely large quantities	（1）无穷小量的概念与性质 （2）无穷小的比较 （3）无穷小的等价代换 （4）无穷大量 （1）concept and properties of infinitesimals （2）comparisions between infinitesimals （3）equivalent transformations of infinitesimals （4）infinitely large quantities	1
1.5	连续函数 continuous function	（1）函数的连续性概念与间断点的分类 （2）连续函数的运算性质与初等函数的连续性 （3）闭区间上连续函数的性质 （4）函数的一致连续性 （1）concept of continuous function and classification of discontinuous points （2）operations on continuous functions and the continuity of elementary functions （3）properties of continuous functions on a closed interval （4）uniform continuity of a function	6

第二章 一元函数微分学及其应用（differential calculus of single variable function and its applications）

章节序号 chapter number	章节名称 chapters	知识点 key points	课时 class hour
2.1	导数 derivatives	（1）导数的概念 （2）导数的几何意义 （3）可导与连续的关系 （4）导数在科学技术中的含义——变化率 （1）concept of derivatives （2）geometric interpretation of derivatives （3）relationship between derivability and continuity （4）meaning of derivative in science and technology——rate of change	3
2.2	求导的基本法则 fundamental derivative rules	（1）函数的和、差、积、商的求导法则 （2）复合函数的求导法则 （3）反函数的求导法则 （4）初等函数的求导法则 （5）高阶导数 （6）隐函数求导法 （7）由参数方程所确定的函数的求导法则 （8）相关变化率问题 （1）derivative rules for sum, difference, product, and quotient of functions （2）derivative rules for composite functions （3）derivative rules for inverse functions （4）derivative rules for elementary functions （5）derivatives of higher order （6）derivative rules of implicit functions （7）derivative rules of a function defined by parametric equations （8）related rates of change	5
2.3	微分 differential	（1）微分的概念 （2）微分的运算法则 （3）高阶微分 （4）微分在近似计算中的应用 （1）concept of the differential （2）rules of operations on differential （3）differential of higher order （4）application of differential in calculating approximation	1

续表

章节序号 chapter number	章节名称 chapters	知识点 key points	课时 class hour
2.4	微分中值定理及其应用 differential mean value theorem and its applications	（1）函数的极值及其必要条件 （2）微分中值定理 （3）洛必达法则 （1）extreme values of a function and its necessary conditions （2）mean value theorem （3）L'Hospital rule	5
2.5	泰勒定理及其应用 Taylor's theorem and its applications	（1）泰勒定理 （2）几个初等函数的麦克劳林公式 （3）泰勒公式的应用 （1）Taylor's theorem （2）Maclaurin formula for some elementary functions （3）applications of Taylor's theorem	2
2.6	函数性态的研究 studies of the behaviors of functions	（1）函数的单调性 （2）函数的极值 （3）函数的最大（小）值 （4）函数图像的凹凸性与拐点 （1）monotonicity of functions （2）extreme values of functions （3）global maxima and minima of a function （4）convexity and inflection points of graphs of functions	5

第三章 一元函数积分学及其应用（integral calculus of single variable functions and its applications）

章节序号 chapter number	章节名称 chapters	知识点 key points	课时 class hour
3.1	定积分的概念、存在条件与性质 concept, existence conditions, and properties of definite integrals	（1）定积分问题举例 （2）定积分的定义 （3）定积分的存在条件 （4）定积分的性质 （1）examples of definite integral problems （2）definition of definite integrals （3）existence conditions for definite integrals （4）properties of definite integrals	3

续表

章节序号 chapter number	章节名称 chapters	知识点 key points	课时 class hour
3.2	微积分基本公式与基本定理 fundamental formulas and theorems of calculus	（1）微积分基本公式 （2）微积分基本定理 （3）不定积分 （1）fundamental formulas of calculus （2）fundamental theorems of calculus （3）indefinite integrals	5
3.3	两种基本积分法 two types of methods of definite integrals	（1）换元积分法 （2）分部积分法 （3）初等函数的积分问题 （1）integration by substitution （2）integration by parts （3）problems of integration for elementary functions	4
3.4	定积分的应用 applications of definite integrals	（1）建立积分表达式的微元法 （2）定积分在几何中的应用举例 （3）定积分在物理中的应用举例 （1）method of elements for setting up integral representations （2）some examples on the applications of the definite integrals in geometry （3）some examples on the applications of the definite integrals in physics	6
3.5	反常积分 improper integrals	（1）无穷区间上的积分 （2）无界函数的积分 （3）无穷区间上的审敛准则 （4）无界函数的审敛准则 （1）integration over an infinite interval （2）integrals of unbounded functions （3）conditions of convergence over an infinite interval （4）conditions of convergence for unbounded functions	7

第四章 常微分方程（ordinary differential equations）

章节序号 chapter number	章节名称 chapters	知识点 key points	课时 class hour
4.1	几类简单的微分方程 some types of simple differential equations	（1）几个基本概念 （2）可分离变量的微分方程 （3）一阶线性微分方程 （4）可用变量代换求解的一阶微分方程 （5）可降阶的高阶微分方程 （6）微分方程应用举例 （1）several basic concepts （2）differential equations with separable variables （3）linear differential equations of first order （4）differential equations of first order that can be solved by variable substitution method （5）reducible differential equations of higher order （6）examples on the applications of differential equations	5
4.2	高阶线性微分方程 linear differential equations of higher order	（1）高阶线性微分方程举例 （2）线性微分方程解的结构 （3）高阶常系数线性齐次微分方程的解法 （4）高阶常系数线性非齐次微分方程的解法 （5）高阶变系数微分的求解问题 （1）examples on application of linear differential equations of higher order （2）structure of solutions for linear differential equations （3）solution of higher order homogeneous linear differential equations with constant coefficients （4）solution of higher order nonhomogeneous linear differential equations with constant coefficients （5）solution of higher order homogeneous linear differential equations with variable coefficients	10
4.3	线性微分方程组 system of linear differential equations	（1）线性微分方程组的基本概念 （2）线性微分方程组解的结构 （3）常系数线性齐次微分方程组的求解方法 （4）常系数线性非齐次微分方程组的求解 （5）微分方程组应用举例 （1）concept of the system of linear differential equations （2）structure of solutions for the system of linear differential equations （3）solution of system of homogeneous linear differential equations with constant coefficients （4）solution of system of nonhomogeneous linear differential equations with constant coefficients （5）example on the applications of system of linear differential equations	6

第五章 多元函数微分学及其应用（multivariable differential calculus and its applications）

章节序号 chapter number	章节名称 chapters	知识点 key points	课时 class hour
5.1	n维欧氏空间中点集的初步知识 preliminary knowledge on points sets in n-dimensional Euclidean space	（1）n维空间$\mathbf{R}^n$ （2）$\mathbf{R}^n$中点列的极限 （3）$\mathbf{R}^n$中的开集与闭集 （4）$\mathbf{R}^n$中的紧集与区域 （1）n-dimensional space $\mathbf{R}^n$ （2）limits of point sequences in $\mathbf{R}^n$ （3）open intervals and closed intervals in $\mathbf{R}^n$ （4）compact sets and regions in $\mathbf{R}^n$	3
5.2	多元函数的极限与连续性 limit and continuity of multivariable functions	（1）多元函数的概念 （2）多元函数的极限与连续性 （3）有界闭区域上多元连续函数的性质 （1）concept of multivariable functions （2）limit and continuity of multivariable functions （3）propertites of multivariable function defined on compact sets	2
5.3	多元数量值函数的导数与微分 partial derivatives and total differentials of multivariable functions	（1）偏导数 （2）全微分 （3）方向导数与梯度 （4）高阶偏导数 （5）多元复合函数的偏导数与全微分 （6）由一个方程确定的隐函数的微分法 （1）partial derivatives （2）total differentials （3）directional derivatives and gradient （4）higher order partial derivatives （5）partial derivatives and total differentials of multivariable composite functions （6）differential of implicit functions defined by one equation	9
5.4	多元函数的泰勒公式与极值问题 Taylor's formula and extreme values for multivariable functions	（1）多元函数的泰勒公式 （2）无约束极值、最大值与最小值 （3）有约束极值和拉格朗日乘数法 （1）Taylor's formula for multivariable functions （2）unconstrained extrema , global maxima and minima （3）extreme values with constraints and the method of Lagrange multipliers	6

续表

章节序号 chapter number	章节名称 chapters	知识点 key points	课时 class hour
5.5	多元向量值函数的导数与微分 derivatives and differentials of vector-valued functions with multiple variables	（1）一元向量值函数的导数与微分 （2）二元向量值函数的导数与微分 （3）微分运算法则 （4）由方程组所确定的隐函数的微分法 （1）derivatives and differentials of vector-valued functions with single variable （2）derivatives and differentials of vector-valued functions with two variables （3）operation rules of differentials （4）differention of implicit functions defined by multiple equations	4
5.6	多元函数微分学在几何上的简单应用 simple applications of multivariable functions in geometry	（1）空间曲线的切线与法平面 （2）弧长 （3）曲面的切平面与法线 （1）tangent line and normal plane to a space curve （2）arc length （3）tangent planes and normal lines of curved surfaces	2
5.7	空间曲线的曲率与挠率 curvature and torsion of space curves	（1）Frenet 标架 （2）曲率 （3）挠率 （1）Frenet frame （2）curvature （3）torsion	4

第六章　多元函数积分学及其应用（multivariable integral calculus and its applications）

章节序号 chapter number	章节名称 chapters	知识点 key points	课时 class hour
6.1	多元数量值函数积分的概念与性质 concept and properties of the integral of a multivariable scalar-valued function	（1）物体质量的计算 （2）多元数量值函数积分的概念 （3）积分存在的条件和性质 （1）computation of the mass of an object （2）concept of the integral of a multivariable scalar-valued function （3）existence conditions and properties of integrals	2

续表

章节序号 chapter number	章节名称 chapters	知识点 key points	课时 class hour
6.2	二重积分的计算 computation of double integrals	（1）二重积分的几何意义 （2）直角坐标系下二重积分的计算方法 （3）极坐标系下二重积分的计算方法 （4）曲线坐标下二重积分的计算方法 （1）geometric meaning of the double integrals （2）computation methods for double integrals in cartesian coordinates （3）computation methods for double integrals in polar coordinates （4）computation methods for double integrals in curvilinear coordinates	4
6.3	三重积分的计算 computation of triple integrals	（1）化三重积分为单积分与二重积分的累次积分 （2）柱面坐标与球面坐标下三重积分的计算法 （1）reduction of a triple integral to an iterated integral consisting of a single integral and a double integral （2）computation of triple integrals in cylindrical and spherical coordinates	2
6.4	含参变量的积分与反常重积分 integral with respect to parameter variables and improper double integrals	（1）含参变量的积分 （2）反常重积分 （1）integral with respect to parameter variables （2）improper double integrals	2
6.5	重积分的应用 applications of multiple integrals	（1）重积分的微元法 （2）应用举例 （1）infinitesimal methods for multiple integrals （2）application examples	2
6.6	第一型线积分与面积分 line integrals and surface integrals of the first type	（1）第一型线积分 （2）第一型面积分 （1）line integrals of the first type （2）surface integrals of the first type	4
6.7	第二型线积分与面积分 line integrals and surface integrals of the second type	（1）场的概念 （2）第二型线积分 （3）第二型面积分 （1）concept of fields （2）line integrals of the second type （3）surface integrals of the second type	8

续表

章节序号 chapter number	章节名称 chapters	知识点 key points	课时 class hour
6.8	各种积分的联系及其在场论中的应用 relations between different kinds of integrals and their application to field theory	（1）格林公式 （2）平面线积分与路径无关的条件 （3）高斯公式与散度 （4）斯托克斯公式与旋度 （5）几种重要的特殊向量场 （1）Green's formula （2）conditions for path independence of planar line integrals （3）Gauss's formula and divergence （4）Stokes's formula and the curl of vectors （5）some important special vector fields	8

第七章 无穷级数（infinite series）

章节序号 chapter number	章节名称 chapters	知识点 key points	课时 class hour
7.1	常数项级数 series with constant terms	（1）常数项级数的概念、性质与收敛原理 （2）正项级数的审敛准则 （3）变号级数的审敛准则 （1）concept，properties and convergence principle of series with constant terms （2）covergence criteria for series with positive terms （3）convergence criteria for alternating series	8
7.2	函数项级数 series of functions	（1）函数项级数的处处收敛性 （2）函数项级数的一致收敛性概念与判别方法 （3）一致收敛级数的性质 （1）everywhere convergence of series of functions （2）concept and criteria of uniform convergence for series of functions （3）propertites of uniformly convergent series	4
7.3	幂级数 power series	（1）幂级数及其收敛半径 （2）幂级数的运算性质 （3）幂级数的应用举例 （1）power series and its radius of convergence （2）operation properties of power series （3）examples on the applications of power series	7

续表

章节序号 chapter number	章节名称 chapters	知识点 key points	课时 class hour
7.4	傅里叶级数 Fourier series	（1）周期函数与三角级数 （2）三角函数系的正交性与傅里叶级数 （3）周期函数的傅里叶展开 （4）定义在[0,l]上函数的傅里叶展开 （1）periodic functions and trigonometric series （2）orthogonality of the system of trigonometric functions and Fourier series （3）Fourier expansion of periodic function （4）Fourier expansion of functions defined on the interval [0,l]	5

3.1.5 实验环节（Experiments）

序号 number	实验内容 experiment content	知识点 key points	课时 class hour
1	基础实验篇 basic experiments	（1）MATLAB软件的基本操作 （2）MATLAB软件绘图 （3）MATLAB软件程序设计 （4）MATLAB软件的基本运算 （5）行列式、矩阵与线性变换 （6）随机数据模拟 （7）无理数的近似计算 （1）basic operations of MATLAB （2）plots in MATLAB （3）program design with MATLAB （4）basic calculations with MATLAB （5）determinant, matrix and linear transformation （6）simulation of random numbers （7）calculat on approximate of irrational numbers	10
2	综合实验篇 comprehensive experiments	（1）线性函数极值求解 （2）非线性函数极值求解 （3）矩阵特征值与迭代法 （4）非线性方程（组）求解 （1）solving extreme values of a linear function （2）solving extreme values of a nonlinear function （3）matrix eigenvalues and iterative methods （4）solving nonlinear equation (sets)	8

续表

序号 number	实验内容 experiment content	知识点 key points	课时 class hour
3	简单建模篇 simple modeling	（1）微分方程模型 （2）数据拟合与数据插值 （3）水塔水流量估计模型 （1）differential equation models （2）data fitting and data interpolations （3）water flowrate estimation model of a water tower	6

3.2 "线性代数与解析几何"教学大纲

课程名称：线性代数与解析几何
Course Title：Linear Algebra and Analytic Geometry
先修课程：初等数学
Prerequisites：Elementary Mathematics
学分：4
Credits：4

3.2.1　课程目的和基本内容（Course Objectives and Basic Contents）

本课程是储能科学与工程专业本科生的基础必修课。本课程的特点是线性代数与空间解析几何融为一体，几何为抽象的代数提供了想象的空间，代数为几何提供了便利的研究工具。代数与几何的融合能够加强学生对数与形内在联系的理解，学会用代数的方法处理几何问题。

本课程与数学分析的内容相互渗透，并为数学分析的多元部分提供必要的代数与几何基础。通过本课程的学习，使学生系统地获取线性代数与空间解析几何的基本知识、基本理论与基本方法，提高运用所学知识分析和解决问题的能力，并为学习相关课程及进一步学习现代数学奠定必要的数学基础。课堂教学中，注重将数学建模思想融入理论课教学，培养学生应用线性代数知识解决实际问题的能力和创新意识。

本课程的主要内容有行列式、向量与矩阵、线性方程组、特征值与特征向量、线性空间与欧氏空间、二次型、几何向量、二次曲面、线性变换等。

This course is a basic compulsory course for undergraduates in energy storage science and engineering.This course features in characterized by the integration of linear algebra with spatial analytic geometry, which provides an imaginary room for abstract algebra and a convenient research tool for geometry. The fusion of algebra and geometry can strengthen students' understanding of the internal relation between numbers and shapes, and help them deal with geometry problems by algebraic method.

This course seeks to integrate linear algebra and analytic geometry together, infiltrate the contents of mathematical analysis, and provide the necessary algebraic and geometric basis for the multivariate part of mathematical analysis. By studying this course, students should systematically acquire some basic knowledge, theory and methods of linear algebra and spatial analytic geometry, improve their ability to analyze and solve problems with the knowledge they have learned, and lay the necessary mathematical foundation for learning related courses and further studying modern mathematics. In classroom teaching, this course takes the thinking of mathematical modelling into theoretical teaching to cultivate students' ability of innovative consciousness and the ability to solve practical problems with linear algebra knowledge.

The main content of this course includes determinant, vector and matrix, system of linear equations, eigenvalues and eigenvectors, linear space and Euclidean space, quadratic form, geometric vector, quadric surface, linear transformation, etc.

3.2.2 课程基本情况（Basic Information of the Course）

课程名称	线性代数与解析几何 Linear Algebra and Analytic Geometry										
开课时间	一年级		二年级		三年级		四年级		总学分	4	总学时	64
	秋	春	秋	春	秋	春	秋	春				
课程定位	储能科学与工程专业本科生数学基础课程群必修课											
授课学时分配	课堂讲授64学时											
先修课程	初等数学											
后续课程	工科数学分析基础、概率论与数理统计, 数学物理方程, 运筹学, 自动控制理论, 电路, 储能系统检测与估计											
教学方式	课堂教学, 综述报告, 讨论, 平时作业, MOOC测验											
考核方式	课程结束笔试成绩占55%, MOOC测验占10%, 平时成绩占5%, 期中考试占30%											
参考教材	[1]李继成, 魏战线.线性代数与解析几何. 3版. 北京:高等教育出版社, 2019.											
参考资料	[1]魏战线.线性代数辅导与典型题解析. 西安:西安交通大学出版社, 2002.											

3.2.3 教学目的和基本要求（Teaching Objectives and Basic Requirements）

（1）掌握工程相关的数学和自然科学知识，理解和领会重要的数学和物理思想方法；

（2）能够应用数学和自然科学的基本原理和方法，建立实际工程问题的合理物理模型，并转化为数学问题进行分析；

（3）针对工程中的具体问题或需求，能够通过文献分析掌握其研究现状及发展趋势，制订研究目标和研究方案，拟订研究技术路线；

（4）根据现代工程技术的发展需求及趋势，了解和掌握解决复杂工程问题所需的现代工具和方法，并理解各自的局限性；

（5）能够开发、选择与使用恰当的技术、资源、现代工程工具和信息技术工具，对工程领域中的复杂工程问题进行模拟、分析和设计，并进行结果验证与评价；

（6）能够合理评价工程实践对环境、社会可持续发展的影响。

（1）Master engineering-related mathematics and natural science knowledge, understand and comprehend important mathematical and physical thinking methods.

（2）Be able to apply basic principles and methods of mathematics and natural sciences to establish reasonable physical models of actual engineering problems, and transform them into mathematical problems for analysis.

（3）For specific problems or needs in the project, students are expected to grasp its research status and development trend through literature analysis, formulate research goals and research plans, and formulate research technical routes.

（4）According to the development needs and trends of modern engineering technology, students should understand and master the modern tools and methods needed to solve complex engineering problems and understand their respective limitations.

（5）Be able to develop, select and use appropriate technologies, resources, modern engineering tools, and information technology tools to simulate, analyze and design complex engineering problems in the engineering field, and to verify and evaluate the results.

（6）Be able to reasonably evaluate the impact of engineering practice on the sustainable development of the environment and society.

3.2.4 课程大纲和知识点（Syllabus and Key Points）

第一章 绪论（introduction）

章节序号 chapter number	章节名称 chapters	知识点 key points	课时 class hour
1.1	行列式的定义与性质 definition and properties of determinant	（1）2阶行列式的定义 （2）n阶行列式的定义 （3）行列式的基本性质 （1）definition of the second-order determinant （2）definition of the n-order determinant （3）basic properties of determinant	2

续表

章节序号 chapter number	章节名称 chapters	知识点 key points	课时 class hour
1.2	行列式的计算 calculation of determinants	（1）行列式的计算 （2）范德蒙德行列式的计算 （1）calculation of determinants （2）calculation of the Vandermonde determinant	2
1.3	克拉默法则 Cramer's rule	克拉默法则 Cramer's rule	2

第二章　矩阵（matrix）

章节序号 chapter number	章节名称 chapters	知识点 key points	课时 class hour
2.1	矩阵及其运算 matrix and its operations	（1）矩阵的概念 （2）矩阵的加法 （3）数与矩阵的乘法 （4）矩阵的乘法 （5）矩阵的转置 （6）方阵的行列式 （1）concept of matrix （2）matrix addition （3）multiplication of numbers and matrices （4）matrix multiplication （5）transpose of a matrix （6）the determinant of a square matrix	2
2.2	逆矩阵 inverse matrix	（1）逆矩阵的概念 （2）方阵可逆的充要条件 （3）逆矩阵的基本性质 （1）concept of inverse matrix （2）necessary and sufficient conditions for invertibility of a square matrix （3）basic properties of the inverse matrix	2
2.3	分块矩阵及其运算 block matrix and its operations	（1）分块矩阵的概念 （2）分块矩阵的运算 （3）分块对角矩阵及其性质 （1）concept of block matrix （2）operations of block matrix （3）block-diagonal matrix and its properties	2

续表

章节序号 chapter number	章节名称 chapters	知识点 key points	课时 class hour
2.4	初等变换与初等矩阵 elementary transformations and elementary matrix	（1）初等变换与初等矩阵的概念 （2）初等变换与初等矩阵的关系 （3）阶梯型矩阵及简化行阶梯型矩阵 （4）用初等变换求可逆矩阵的逆矩阵 （1）concept of elementary transformation and elementary matrix （2）relationship between elementary transformations and elementary matrices （3）matrices of echelon form and reduced echelon form （4）using elementary transformations to find the inverse matrix of invertible matrix	2
2.5	矩阵的秩 rank of matrix	（1）矩阵的秩的概念 （2）计算矩阵的秩的一般方法 （1）concept of the rank of matrix （2）general method for calculating the rank of matrix	2

第三章 几何向量及其应用 (geometric vectors and their applications)

章节序号 chapter number	章节名称 chapters	知识点 key points	课时 class hour
3.1	向量及其线性运算 vectors and their linear operations	（1）向量的基本概念 （2）向量共线、共面的充要条件 （3）直角坐标系和向量的坐标、长度和方向余弦 （4）用坐标进行向量的线性运算、正交射影 （1）basic concept of vectors （2）necessary and sufficient conditions for vectors to be collinear and coplanar （3）coordinate, length, and directional cosine of cartesian coordinate system and vectors （4）linear operation and orthogonal projection of vectors with coordinates	2
3.2	数量积、向量积、混合积 dot product、vector product、mixed product	（1）数量积 （2）向量积 （3）混合积 （1）dot product （2）vector product （3）mixed product	2

续表

章节序号 chapter number	章节名称 chapters	知识点 key points	课时 class hour
3.3	平面和空间直线 planes and lines in space	（1）平面方程 （2）直线方程 （3）两条直线的位置关系 （4）两个平面的位置关系 （1）equations of a plane （2）equations of a line （3）relative positions of two lines （4）relative positions of two planes	4

第四章 n 维向量与线性方程组（n-dimensional vectors and systems of linear equations）

章节序号 chapter number	章节名称 chapters	知识点 key points	课时 class hour
4.1	消元法 elimination method	（1）消元法 （2）线性方程组的解 （3）线性方程组有解判定定理 （1）elimination method （2）solutions of system of linear equations （3）theorems on the existence solutions of system of linear equations	2
4.2	向量组的线性相关性 linear dependence of a set of vectors	（1）n维向量及其线性运算 （2）线性组合与线性表示 （3）等价向量组 （4）线性相关与线性无关 （1）n-dimensional vectors and their linear operation （2）linear combination and linear representation （3）equivalent sets of vectors （4）linear dependence and linear independence	4
4.3	向量组的秩 rank of a set of vectors	（1）向量组的极大无关组 （2）向量组的秩 （3）向量组的秩与矩阵的秩的关系 （1）maximal independent system of a set of vectors （2）rank of a set of vectors （3）relationship between the rank of a set of vectors and the rank of a matrix	3

章节序号 chapter number	章节名称 chapters	知识点 key points	课时 class hour
4.4	线性方程组的解的结构 structure of the solution to a system of linear equations	（1）齐次线性方程组的解的性质和基础解系 （2）非齐次线性方程组的解的性质和结构 （1）properties of solutions of system of homogeneous linear equations and basic solutions （2）properties and structures of solutions of nonhomogeneous system of linear equations	3

第五章　线性空间与欧氏空间（linear space and Euclidean space）

章节序号 chapter number	章节名称 chapters	知识点 key points	课时 class hour
5.1	线性空间的基本概念 basic concepts of linear space	（1）线性空间的定义 （2）线性空间的基本性质 （3）线性子空间的定义 （4）基、维数和向量的坐标 （5）基变换与坐标变换 （6）线性空间的同构 （7）子空间的交与和 （1）definition of linear space （2）basic properties of linear space （3）definition of linear subspace （4）basis, dimension, coordinates of a vector （5）change of bases and coordinate transformation （6）iso morphic linear spaces （7）intersection and sum of subspaces	4
5.2	欧氏空间的基本概念 basic concepts of Euclidean space	（1）内积及其基本性质 （2）范数和夹角 （3）标准正交基及其基本性质 （4）格拉姆-施密特正交化方法 （5）正交矩阵 （1）inner product and its basic properties （2）norm and included angle （3）orthonormal basis and its basic properties （4）Gram-Schmidt orthogonalization method （5）orthogonal matrix	4

第六章 特征值与特征向量（eigenvalues and eigenvectors）

章节序号 chapter number	章节名称 chapters	知识点 key points	课时 class hour
6.1	矩阵的特征值与特征向量 eigenvalues and eigenvectors of a matrix	（1）特征值与特征向量的概念 （2）特征方程、特征多项式与特征子空间 （3）求特征值与特征向量的一般步骤 （4）特征值和特征向量的性质 （5）特征值和特征向量的计算 （1）concepts of eigenvalues and eigenvectors （2）characteristic equations, characteristic polynomials and eigenspaces （3）general steps to evaluate eigenvalues and eigenvectors （4）properties of eigenvalues and eigenvectors （5）calculations of eigenvalues and eigenvectors	3
6.2	相似矩阵与矩阵的相似对角化 similar matrices and similarity to a diagonal matrix	（1）相似矩阵 （2）矩阵可对角化的条件 （3）实对称矩阵的对角化 （1）similar matrices （2）conditions for diagonalizable （3）diagonalization of real symmetric matrices	3

第七章 二次曲面与二次型（quadratic surfaces and quadratic forms）

章节序号 chapter number	章节名称 chapters	知识点 key points	课时 class hour
7.1	曲面与空间曲线 surfaces and curves in space	（1）曲面与空间曲线的方程 （2）柱面、锥面、旋转面 （3）五种典型的二次曲面 （4）曲线在坐标面上的投影 （5）空间区域的简图 （1）equations of surfaces and curves in space （2）cylinder, cone and rotation surface （3）five typical quadratic surfaces （4）projection of the curve on the coordinate surface （5）schematic diagram of spatial region	3

续表

章节序号 chapter number	章节名称 chapters	知识点 key points	课时 class hour
7.2	实二次型 quadratic form	（1）实二次型及其矩阵表示 （2）实二次型的标准形 （3）合同变换与惯性定理 （4）正定实二次型 （1）quadratic form and its matrix representation （2）standard form of a quadratic form （3）congruence transformation and an inertia theorem （4）positive definite quadratic form	5

第八章　线性变换（linear transformations）

章节序号 chapter number	章节名称 chapters	知识点 key points	课时 class hour
8.1	线性变换及其运算 linear transformations and their operations	（1）线性变换的定义与性质 （2）线性变换的核与值域 （3）线性变换的基本定理 （4）线性变换的运算 （1）definition and properties of linear transformations （2）kernel and range of linear transformations （3）basic theorems of linear transformations （4）operations of linear transformations	3
8.2	线性变换的矩阵表示 matrix representation of linear transformations	（1）线性变换的矩阵 （2）线性变换的矩阵举例 （3）线性算子在不同基下矩阵的关系 （1）matrix of linear transformations （2）examples on the matrices of linear transformations （3）relations between matrices of linear operators on different bases	3

3.3　"概率论与数理统计"教学大纲

课程名称：概率论与数理统计

Course Title：Probability Theory and Mathematical Statistics

先修课程：高等数学Ⅰ，线性代数与解析几何

Prerequisites：Advanced Mathematics Ⅰ，Linear Algebra and Analytic Geometry

学分：3

Credits：3

3.3.1 课程目的和基本内容（Course Objectives and Basic Contents）

本课程是储能科学与工程专业本科生的基础必修课，也是工科学生必修的一门数学基础课，是研究随机现象的统计规律性的一门课程，在工程技术中有很广泛的应用。学习和正确运用概率统计方法并解决工程问题已成为工科类大学生的基本要求。通过本课程的学习，能使学生掌握概率论与数理统计的基本思想和基本方法，提高学生运用概率统计知识分析和解决实际问题的能力，并为学习后继课程和继续深造打好基础。

本课程涵盖概率论与数理统计两部分内容，包括随机事件与概率、一维随机变量及其分布、多维随机变量及其分布、随机变量的数字特征、大数定律和中心极限定理、数理统计的基本概念、参数估计、假设检验等。

This course is a basic mandatory mathematics course for engineering students, which studies statistical laws of random phenomena and has a wide range of applications in engineering technology. It has become a basic requirement for engineering students to learn and correctly use probability theory and mathematical statistics methods to solve engineering problems. By studying this course, students could master the basic ideas and methods of probability theory and mathematical statistics, and improve the ability to analyze and solve practical problems using these methods. Furthermore, it can lay a good foundation for subsequent courses and further studies.

This course focuses on the probability theory and mathematical statistics, which includes random events and probability, one-dimensional random variables and their distributions, multiple random variables and their distributions, numerical characteristics of random variables, the law of large numbers and the central limit theorem, basic concepts of mathematical statistics, parameter estimation, hypothesis testing, etc.

3.3.2 课程基本情况（Basic Information of the Course）

课程名称	概率论与数理统计 Probability Theory and Mathematical Statistics										
开课时间	一年级		二年级		三年级		四年级		总学分	3	总学时	48
	秋	春	秋	春	秋	春	秋	春				
课程定位	储能科学与工程专业本科生数学基础课程群必修课											
授课学时分配	课堂讲授48学时											

先修课程	高等数学Ⅰ,线性代数与解析几何
后续课程	储能系统检测与估计
教学方式	课堂教学,平时作业,MOOC课程学习,讨论
考核方式	课程结束笔试成绩占60%,期中考试占30%,MOOC测试占10%
参考教材	[1] 王宁,王峰,施雨.概率统计与随机过程.2版.西安:西安交通大学出版社,2019.
参考资料	[1] 施雨,李耀武.概率论与数理统计应用.2版.西安:西安交通大学出版社,2005. [2] 魏平,王宁.概率统计辅导书.西安:西安交通大学出版社,2010.

3.3.3 教学目的和基本要求（Teaching Objectives and Basic Requirements）

（1）掌握概率论与数理统计的基本思想和基本方法；

（2）学会用随机量去认识客观世界中的随机现象，提高学生运用概率统计知识分析和解决实际问题的能力；

（3）通过课后自主实践，加强学生对随机模型的辨识能力，提高学生的实践意识和创新能力；

（4）培养学生正确认识随机量的思维能力，强化随机数学素养；

（5）拓展学生的知识结构，培养学生自主学习和不断学习的意识和能力。

（1）Master the basic philosophies and methods of probability theory and mathematical statistics.

（2）Learn to recognize random phenomena in the real world with random quantity, and improve students'ability to analyze and solve practical problems with knowledge of probability and statistics.

（3）Strengthen the identification ability of the random model through students' independent practice after class, and improve students' practical awareness and innovation ability.

（4）Train students' thinking ability to correctly understand the random quantity and strengthen random mathematics literacy.

（5）Expand students' knowledge structure and cultivate students' awareness and ability of independent learning and continuous learning.

3.3.4 课程大纲和知识点（Syllabus and Key Points）

第一章 随机事件与概率（random events and probability）

章节序号 chapter number	章节名称 chapters	知识点 key points	课时 class hour
1.1	随机事件 random events	（1）随机现象与随机试验 （2）样本空间与随机事件 （3）事件的关系与运算 （1）random phenomena and random trials （2）sample spaces and random events （3）relations and operations of events	2
1.2	概率 probability	（1）概率的古典定义 （2）概率的统计定义 （3）概率的公理化定义 （4）概率的性质 （1）classical definition of probability （2）statistical definition of probability （3）axiomatic definition of probability （4）properties of probability	2
1.3	条件概率、事件的相互独立性 conditional probability，mutual independence of events	（1）条件概率与乘法公式 （2）全概率公式与贝叶斯公式 （3）事件的相互独立性 （1）conditional probability and multiplication formula （2）total probability formula and Bayes' formula （3）mutual independence of events	3

第二章 一维随机变量及其分布（one dimensional random variable and its distributions）

章节序号 chapter number	章节名称 chapters	知识点 key points	课时 class hour
2.1	一维随机变量 one dimensional random variables	（1）随机变量与分布函数 （2）离散型随机变量 （3）连续型随机变量 （1）random variables and distribution functions （2）discrete random variables （3）continuous random variables	4
2.2	一维随机变量函数的概率分布 the probability distribution of functions with one dimensional random variable	（1）一维随机变量函数 （2）一维离散型随机变量函数的概率分布 （3）一维连续型随机变量函数的概率分布 （1）function of one dimensional random variable （2）probability distribution of function with one dimensional discrete random variable （3）probability distribution of function with one dimensional continuous random variable	1

第三章　多维随机变量及其分布（multiple dimensional random variables and their distributions）

章节序号 chapter number	章节名称 chapters	知识点 key points	课时 class hour
3.1	n维随机变量 n-dimensional random variables	（1）n维随机变量的概念 （2）分布函数与边缘分布函数 （3）二维离散型随机变量 （4）二维连续型随机变量 （1）concept of n-dimensional random variables （2）distribution functions and marginal distribution functions （3）two-dimensional discrete random variables （4）two-dimensional continuous random variables	3
3.2	条件分布 conditional distribution	（1）二维离散型随机变量的条件分布 （2）二维连续型随机变量的条件分布 （1）conditional distribution of two-dimensional discrete random variables （2）conditional distribution of two-dimensional continuous random variables	1
3.3	随机变量的相互独立性 mutual independence of random variables	（1）二维离散型随机变量的相互独立性 （2）二维连续型随机变量的相互独立性 （1）mutual independence of two-dimensional discrete random variables （2）mutual independence of two-dimensional continuous random variables	1
3.4	二维随机变量函数的概率分布 probability distribution of functions with two-dimensional random variables	（1）多维随机变量函数 （2）二维离散型随机变量函数的概率分布 （3）二维连续型随机变量函数的概率分布 （1）functions of n-dimensional random variables （2）probability distribution of functions with two-dimensional discrete random variables （3）probability distribution of functions with two-dimensional continuous random variables	4

第四章 随机变量的数字特征（numerical characteristics of random variables）

章节序号 chapter number	章节名称 chapters	知识点 key points	课时 class hour
4.1	数学期望 mathematical expectation	（1）数学期望的概念 （2）随机变量函数的数学期望 （3）数学期望的性质 （1）concept of mathematical expectation （2）mathematical expectation of random variable functions （3）properties of mathematical expectation	2
4.2	方差 variance	（1）方差和标准差的概念 （2）方差的性质 （1）concept of variance and standard deviation （2）properties of variance	2
4.3	协方差与相关系数 covariance and correlation coefficient	（1）协方差与相关系数 （2）矩 （1）covariance and correlation coefficient （2）moment	2

第五章 大数定律与中心极限定理（law of large numbers and central limit theorem）

章节序号 chapter number	章节名称 chapters	知识点 key points	课时 class hour
5.1	大数定律 law of large numbers	（1）切比雪夫不等式 （2）大数定律 （1）Chebyshev's inequality （2）law of large numbers	1
5.2	中心极限定理 central limit theorem	（1）独立同分布中心极限定理 （2）棣莫弗-拉普拉斯中心极限定理 （1）central limit theorem for independent identically distribution （2）De Moivre-Laplace central limit theorem	2

第六章　数理统计的基本概念（basic concept of mathematical statistics）

章节序号 chapter number	章节名称 chapters	知识点 key points	课时 class hour
6.1	总体与样本 population and samples	（1）总体及其分布 （2）样本 （1）population and its distribution （2）samples	1
6.2	样本分布 sample distribution	（1）样本频率分布 （2）频率直方图 （3）经验分布函数 （1）sample frequency distribution （2）frequency histogram （3）empirical distribution function	1
6.3	统计量 statistics	（1）统计量的概念 （2）几个常用的统计量 （1）concept of statistics （2）several commonly used statistics	1
6.4	抽样分布 sampling distribution	（1）几个常用的重要分布 （2）分位数 （3）正态总体的抽样分布 （1）several important distributions （2）quantile （3）sampling distribution of normal population	3

第七章　参数估计（parameter estimation）

章节序号 chapter number	章节名称 chapters	知识点 key points	课时 class hour
7.1	点估计 point estimation	（1）矩估计 （2）最大似然估计 （1）moment estimation （2）maximum likelihood estimation	2
7.2	估计量的评选标准 criteria for evaluation of estimators	（1）无偏性 （2）有效性 （3）相合性 （1）unbiased （2）efficiency （3）consistence	1

章节序号 chapter number	章节名称 chapters	知识点 key points	课时 class hour
7.3	区间估计 interval estimation	（1）双侧区间估计 （2）单侧区间估计 （1）two-sided interval estimation （2）one-sided interval estimation	1
7.4	正态总体参数的区间估计 interval estimation of normal population parameters	（1）单个总体的情形 （2）两个总体的情形 （1）case of a single normal population （2）case of two normal populations	2

第八章 假设检验（hypothesis testing）

章节序号 chapter number	章节名称 chapters	知识点 key points	课时 class hour
8.1	假设检验的基本概念 basic concept of hypothesis testing	（1）假设检验的基本原理 （2）两类错误 （3）假设检验的一般步骤 （1）basic principles of hypothesis testing （2）two types of errors （3）general steps of hypothesis testing	2
8.2	正态总体参数的假设检验 hypothesis testing of normal population parameters	（1）单个总体的情形 （2）两个总体的情形 （1）case of a single normal population （2）case of two normal populations	1.5
8.3	单边假设检验 one-sided hypothesis testing	单边假设检验 one-sided hypothesis testing	0.5
8.4	大样本检验及成对数据检验 large sample testing and paired data testing	（1）非正态总体的大样本检验 （2）成对数据假设检验 （1）large sample testing of abnormal population （2）paired data testing	0.5
8.5	分布假设检验 testing of distribution hypothesis	（1）分布拟合检验 （2）卡方拟合检验法 （1）distribution goodness of fit testing （2）chi-square goodness of fit testing	0.5

第九章　回归分析（regression analysis）

章节序号 chapter number	章节名称 chapters	知识点 key points	课时 class hour
9.1	一元线性回归 simple linear regression	（1）一元线性回归模型 （2）参数的估计 （3）回归方程的检验 （4）利用回归方程进行预测 （5）可化为一元线性回归的一元非线性回归 （1）simple linear regression model （2）estimation of parameters （3）test of regression equation （4）prediction by regression equation （5）simple nonlinear regression which can be reduced to simple linear regression	1

3.4　"复变函数与积分变换"教学大纲

课程名称：复变函数与积分变换
Course Title：Functions of Complex Variable and Integral Transforms
先修课程：高等数学 I
Prerequisites：Advanced Mathematics I
学分：3
Credits：3

3.4.1　课程目的和基本内容（Course Objectives and Basic Contents）

本课程是储能科学与工程专业本科生的基础必修课，培养学生的数学素质，提高其数学认知能力和应用数学知识解决实际问题的能力。

本课程的内容包括复数与复变函数，复变函数的积分及其性质，解析函数的性质（包括高阶导数公式），幂级数展开，孤立奇点的分类（包括无穷远点），留数及其应用，共形映射的概念及性质（特别要掌握分式线性映射以及几个初等函数定义的映射所具有的性质，傅里叶变换和拉普拉斯变换的定义、性质及其应用。

This course is a fundamental compulsory course for energy storage science and engineering students. The purpose is to cultivate mathematical cognitive ability and to apply mathematical knowledge to solve practical problems.

The course focuses on complex numbers and function of complex variable, integral of function of complex variable and their properties, properties of analytic functions (including the higher derivative formula), power series expansion, classification of isolated singularities (including infinity), residue and its applications, concept and properties of conformal mappings (fractional linear mappings and some elementary function mappings), definitions, properties and applications of the Fourier transform and the Laplace transform.

3.4.2 课程基本情况（Basic Information of the Course）

课程名称	复变函数与积分变换 Functions of Complex Variable and Integral Transforms							
开课时间	一年级	二年级	三年级	四年级	总学分	3	总学时	48
	秋　春	秋　春	秋　春	秋　春				
课程定位	储能科学与工程专业本科生数学基础课程群必修课							
授课学时分配	课堂讲授48学时							
先修课程	高等数学 I							
后续课程	自动控制理论							
教学方式	课堂教学,讨论,作业,结合MOOC混合式教学							
考核方式	课程结束笔试成绩占70%,作业占10%,MOOC测验占20%							
参考教材	[1] 王绵森.复变函数. 2版. 北京:高等教育出版社,2020. [2] 张元林.积分变换. 6版. 北京:高等教育出版社,2019.							
参考资料	[1] 王绵森.复变函数学习辅导与习题选解. 北京:高等教育出版社,2003. [2] 张元林.积分变换习题全解指南. 北京:高等教育出版社,2004.							

3.4.3 教学目的和基本要求（Teaching Objectives and Basic Requirements）

（1）掌握复变函数的数学理论体系和重要思想方法，提高运用所学知识分析和解决实际问题的能力；

（2）熟悉基本概念和定理的几何背景和实际应用背景，强调对课程内容知识的本质理解和实际工程应用；

（3）通过共形映射的思想方法，提高创新意识和创新能力；

（4）培养严谨的数学逻辑思维能力，提升编程计算能力，强化数学素养；

（5）理解积分变换的思想，掌握积分变换的性质，并加以熟练应用，拓展知识面，为后继课程打好基础。

（1）Understand the mathematical theory system and important ideology of functions of complex variable and integral transforms, improve the ability to use the knowledge to analyze and solve practical problems.

（2）Be familiar with the geometric and practical background of basic concepts and theorems, with the highlight of a thorough understanding of core contents and applications in engineering.

（3）Improve the innovation ability from the ideology of conformal mapping.

（4）Improve the logical thinking and the numerical programming abilities, enhance the mathematics accomplishment.

（5）Understand the ideology and properties of integral transformation, apply them to problem-solving, and set up a solid foundation for subsequent courses.

3.4.4　课程大纲和知识点（Syllabus and Key Points）

第一章　复数与复变函数（complex numbers and functions of complex variable）

章节序号 chapter number	章节名称 chapters	知识点 key points	课时 class hour
1.1	复数的概念与运算 concepts and operations of complex numbers	（1）复数及其代数运算 （2）复数的几何表示 （3）复数的乘幂与方根 （4）复数在几何上的应用举例 （5）复球面与无穷远点 （1）complex numbers and their algebraic operations （2）geometric representation of complex numbers （3）powers and square root of the complex numbers （4）examples of applications of complex numbers in geometry （5）complex sphere and infinity point	1
1.2	复变函数及其极限与连续性 limits and continuity of functions of complex variable	（1）复平面上的区域 （2）复变函数的概念 （3）复变函数的极限与连续性 （1）regions on the complex plane （2）concept of functions of complex variable （3）limit and continuity of functions of complex variable	2

第二章 解析函数(analytic functions)

章节序号 chapter number	章节名称 chapters	知识点 key points	课时 class hour
2.1	解析函数的概念及其判定 concepts and determination of analytic functions	(1)复变函数的导数与微分 (2)解析函数的概念 (3)判定函数解析性的方法 (1)derivatives and differentials of functions of complex variable (2)concept of analytic functions (3)method to determine an analytic function	2
2.2	复变初等函数 complex elementary functions	(1)指数函数 (2)对数函数 (3)乘幂与幂函数 (4)三角函数与双曲函数 (5)反三角函数与反双曲函数 (1)exponential functions (2)logarithmic functions (3)powers and power functions (4)trigonometric functions and hyperbolic functions (5)inverse trigonometric functions and inverse hyperbolic functions	2

第三章 复变函数的积分(integrals of functions of complex variable)

章节序号 chapter number	章节名称 chapters	知识点 key points	课时 class hour
3.1	复变函数积分的概念、性质及其计算 concepts, properties, and calculation of integrals of functions of complex variable	(1)积分的定义 (2)积分的存在条件与计算方法 (3)积分的基本性质 (1)definition of integral (2)existence condition and calculation method of integral (3)basic properties of integrals	1
3.2	柯西-古萨定理及其推广 Cauchy-Goursat theorem and its generalization	(1)柯西-古萨基本定理 (2)柯西-古萨定理的推广——复合闭路定理 (1)Cauchy- Goursat fundamental theorem (2)generalization of the Cauchy- Goursat theorem——composite closed circuit theorem	1

续表

章节序号 chapter number	章节名称 chapters	知识点 key points	课时 class hour
3.3	原函数与不定积分 antiderivative functons and indefinite integrals	（1）原函数 （2）不定积分 （1）antiderivative functions （2）indefinite integrals	1
3.4	柯西积分公式与高阶导数公式 Cauchy integral formula and higher-order derivative formula	（1）柯西积分公式 （2）高阶导数公式与解析函数的无限可微性 （1）Cauchy integral formula （2）higher-order derivative formulas and infinite differentiability of analytic functions	1
3.5	解析函数与调和函数的关系 relationship between analytic and harmonic functions	（1）调和函数的概念 （2）解析函数与调和函数的关系 （1）concept of harmonic functions （2）relationship between analytic functions and harmonic functions	1

第四章　复变函数项级数（series of complex fuctions）

章节序号 chapter number	章节名称 chapters	知识点 key points	课时 class hour
4.1	复数项级数与复变函数项级数 series of complex terms and series of functions of complex variable	（1）复数列的概念 （2）复数项级数 （3）复变函数项级数 （1）concept of complex sequences （2）series of complex numbers （3）series of functions of complex variable	2
4.2	幂级数 power series	（1）幂级数的收敛性 （2）幂级数的收敛圆与收敛半径 （3）幂级数的运算性质 （1）convergence of power series （2）convergence disk and convergence radius of the power series （3）operation properties of power series	2
4.3	泰勒级数 Taylor series	（1）解析函数的泰勒展开定理 （2）求解析函数泰勒展开的方法 （1）Taylor expansion theorem of analytic functions （2）method of Taylor expansion of analytic function	2

续表

章节序号 chapter number	章节名称 chapters	知识点 key points	课时 class hour
4.4	洛朗级数 Laurent series	（1）解析函数的洛朗展开定理 （2）求解析函数洛朗展开的方法 （1）Laurent expansion theorem of analytic functions （2）method of Laurent expansion of analytic function	2

第五章　留数及其应用（residue and its applications）

章节序号 chapter number	章节名称 chapters	知识点 key points	课时 class hour
5.1	解析函数的孤立奇点 isolated singularities of analytic functions	（1）孤立奇点及其分类 （2）函数的零点与极点的关系 （3）函数在无穷远点的性态 （1）isolated singularities and their classification （2）relationship between the zero point and the pole of the function （3）properties of the function at infinity	2
5.2	留数与留数定理 residue and theorem of residues	（1）留数的定义及留数定理 （2）计算留数的方法 （1）definition of residue and theorem of residues （2）method of computing residue	2
5.3	留数定理在计算实积分中的应用 application of the theorem of residues in calculating real integrals	（1）形如 $\int_0^{2\pi} R(\cos\theta,\sin\theta)\mathrm{d}\theta$ 的积分 （2）形如 $\int_{-\infty}^{+\infty} R(x)\mathrm{d}x$ 的积分 （3）形如 $\int_{-\infty}^{+\infty} R(x)\mathrm{e}^{aix}\mathrm{d}x(a>0)$ 的积分 （1）integral of the form $\int_0^{2\pi} R(\cos\theta,\sin\theta)\mathrm{d}\theta$ （2）integral of the form $\int_{-\infty}^{+\infty} R(x)\mathrm{d}x$ （3）integral of the form $\int_{-\infty}^{+\infty} R(x)\mathrm{e}^{aix}\mathrm{d}x(a>0)$	2

第六章　共形映射（conformal mapping）

章节序号 chapter number	章节名称 chapters	知识点 key points	课时 class hour
6.1	共形映射的概念 concept of conformal mapping	（1）解析函数导数的几何意义 （2）共形映射的概念与单叶解析函数的共形映射 （1）geometric meaning of derivative of analytic function （2）concept of conformal mapping and the conformal mapping of single-leaf analytic functions	2

续表

章节序号 chapter number	章节名称 chapters	知识点 key points	课时 class hour
6.2	分式线性映射 fractional linear mapping	（1）分式线性映射及其构成 （2）分式线性映射的性质 （3）分式线性映射应用举例 （1）fractional linear mapping and its composition （2）properties of fractional linear mapping （3）application example of fractional linear mapping	2
6.3	几个初等函数所构成的 共形映射 conformal mapping of several elementary functions	（1）幂函数与根式函数 （2）指数函数与对数函数 （1）power function and root function （2）exponential function and logarithmic function	2

第七章　傅里叶变换（Fourier transform）

章节序号 chapter number	章节名称 chapters	知识点 key points	课时 class hour
7.1	傅里叶积分 Fourier integral	傅里叶积分的定义 definition of the Fourier integral	1
7.2	傅里叶变换 Fourier transform	（1）傅里叶变换的概念 （2）单位脉冲函数及其傅里叶变换 （3）非周期函数的频谱 （1）concept of the Fourier transform （2）unit impulse function and its Fourier transform （3）spectrum of aperiodic functions	1
7.3	傅里叶变换的性质 properties of the Fourier transform	（1）线性性质 （2）位移性质 （3）微分性质 （4）积分性质 （1）linear properties （2）displacement properties （3）differential properties （4）integral properties	3
7.4	卷积 convolution	（1）卷积定义 （2）卷积定理 （1）definition of convolution （2）convolution theorem	2

续表

章节序号 chapter number	章节名称 chapters	知识点 key points	课时 class hour
7.5	傅里叶变换的应用 applications of the Fourier transform	（1）利用傅里叶变换求解常微分方程 （2）利用傅里叶变换求解积分方程 （1）use the Fourier transform to solve ordinary differential equations （2）use the Fourier transform to solve integral equation	1

第八章 拉普拉斯变换（Laplace transform）

章节序号 chapter number	章节名称 chapters	知识点 key points	课时 class hour
8.1	拉普拉斯变换的概念 concept of the Laplace transform	（1）拉普拉斯变换的定义 （2）拉普拉斯变换存在定理 （1）definition of the Laplace transform （2）existence theorem of the Laplace transform	1
8.2	拉普拉斯变换的性质 properties of the Laplace transform	（1）线性性质 （2）微分性质 （3）积分性质 （4）位移性质 （5）延迟性质 （1）linear property （2）differential property （3）integral property （4）displacement property （5）time shifting property	4
8.3	卷积 convolution	（1）卷积的概念 （2）卷积定理 （1）concept of convolution （2）convolution theorem	1
8.4	拉普拉斯逆变换 inverse transform of the Laplace transform	逆变换定理 inverse transform theorem	1
8.5	拉普拉斯变换的应用 applications of the Laplace transform	（1）利用拉普拉斯变换求解常微分方程 （2）利用拉普拉斯变换解积分方程 （1）using Laplace transform to solve ordinary differential equations （2）using Laplace transform to solve the integral equations	1

3.5 "数学物理方程"教学大纲

课程名称：数学物理方程

Course Title：Mathematical Physics with Partial Differential Equations

先修课程：工科数学分析，线性代数与解析几何

Prerequisites：Mathematical Analysis in Engineering, Linear Algebra and Analytic Geometry

学分：2

Credits：2

3.5.1 课程目的和基本内容（Course Objectives and Basic Contents）

本课程是储能科学与工程专业本科生的基础必修课,关注三类典型二阶偏微分方程(波动方程、热传导方程和拉普拉斯方程)，介绍方程相关的基本概念和求解算法。主要介绍三类方程的建立和分离变量法，同时介绍格林函数法和方程特征线法。

This course is a basic compulsory course for undergraduates in energy storage science and engineering.The course focuses on the three prototypical second-order PDEs——the wave equation, heat conduction equation, and Laplace's equation. As a basic course in engineering, it is designed to present the related concepts and solving algorithms for the PDEs at an introductory level. This course introduces the derivation of these equations, and separation of variables in detail. It also introduces Green's functions and characteristic curves, and how these methods are used to solve PDEs.

3.5.2 课程基本情况（Basic Information of the Course）

课程名称	数学物理方程 Mathematical Physics with Partial Differential Equations											
开课时间	一年级		二年级		三年级		四年级		总学分	2	总学时	32
	秋	春	秋	春	秋	春	秋	春				
课程定位	储能科学与工程专业本科生数学基础课程群必修课											
授课学时分配	课堂讲授32学时											
先修课程	高等数学Ⅰ,线性代数与解析几何											
后续课程	无											
教学方式	课堂教学,讨论,作业											
考核方式	课程结束笔试成绩占80%,实验报告占0%,平时成绩占20%											
参考教材	[1]申建中,刘峰.数学物理方程.2版.西安:西安交通大学出版社,2018.											
参考资料	[1]姚端正.数学物理方法学习指导.北京:科学出版社,2003.											

3.5.3 教学目的和基本要求（Teaching Objectives and Basic Requirements）

（1）培养学生应用 PDEs 解决实际问题的能力。学生应初步掌握三类典型方程的建立方法并理解偏微分方程的相关概念。

（2）培养学生应用分离变量法求解方程的能力。学生应熟悉三类方程的分离变量法，理解贝塞尔函数。

（3）培养学生选择或设计有效算法的能力。学生应掌握格林函数法和特征线法。

（1）Cultivate students' ability to analyze and solve practical problems using PDEs. Students should handle the establishment of the three typical second-order PDEs, and comprehend concepts related to them.

（2）Cultivate students' ability to solve PDEs using separation of variables. Students should be proficient in separation of variables for the three prototypical second-order PDEs, and comprehend Bessel functions.

（3）Cultivate students' ability to select or design effective algorithm for PDEs. Students should handle Green's functions and characteristic curves, and be able to solve PDEs with them.

3.5.4 课程大纲和知识点（Syllabus and Key Points）

第一章 绪论（introduction）

章节序号 chapter number	章节名称 chapters	知识点 key points	课时 class hour
1.1	数学模型的建立 establishment of mathematical models	（1）波动方程 （2）热传导方程 （3）泊松方程 （4）初始条件和边界条件 （1）wave equation （2）heat conduction equation （3）Poisson's equation （4）initial conditions and boundary conditions	4
1.2	定解问题的适定性 well-posedness for problems with definite solution	（1）适定性 （2）古典解 （1）well-posedness （2）classical solution	2
1.3	叠加原理 superposition principle	（1）叠加原理 （2）边界条件齐次化 （1）superposition principle （2）boundary condition homogenization	2

第二章　分离变量法（methods based on separation of variables）

章节序号 chapter number	章节名称 chapters	知识点 key points	课时 class hour
2.1	特征值问题 eigenvalue problems	（1）特征值 （2）特征函数 （1）eigenvalues （2）eigenfunctions	2
2.2	分离变量法 separation of variables	（1）弦振动方程 （2）热传导方程 （3）拉普拉斯方程 （1）string vibration equations （2）heat conduction equations （3）Laplace's equation	6

第三章　贝塞尔函数（Bessel's function）

章节序号 chapter number	章节名称 chapters	知识点 key points	课时 class hour
3.1	贝塞尔函数 Bessel's function	（1）Γ-函数 （2）贝塞尔方程 （3）贝塞尔函数的性质 （4）贝塞尔级数 （1）Γ-function （2）Bessel equation （3）property of Bessel function （4）Bessel series	6
3.2	多个自变量分离变量法举例 separation of multiple variables	（1）圆域上波动方程 （2）圆域上热传导方程 （1）wave equations on a disk （2）heat conduction equation on a disk	2

第四章　格林函数（Green Function）

章节序号 chapter number	章节名称 chapters	知识点 key points	课时 class hour
4.1	格林公式 Green formula	格林公式 Green formula	2

续表

章节序号 chapter number	章节名称 chapters	知识点 key points	课时 class hour
4.2	拉普拉斯方程基本解和格林函数 fundamental solution of Laplace equation and Green function	（1）基本解 （2）格林函数 （1）fundamental solution （2）Green function	1
4.3	狄利克雷问题 Dirichlet problem	格林函数构造 construction of Green function	1

第五章 特征线法（characteristic line）

章节序号 chapter number	章节名称 chapters	知识点 key points	课时 class hour
5.1	一阶偏微分方程特征线法 characteristic line method for first-order partial differential equations	（1）特征方程 （2）特征线 （1）characteristic equation （2）characteristic curve	2
5.2	一维波动方程特征线法 characteristic line method for one-dimensional wave equation	行波解 traveling wave solution	2

3.6 "运筹学"教学大纲

课程名称：运筹学

Course Title：Operations Research

先修课程：高等数学Ⅰ，线性代数与解析几何

Prerequisites：Advanced Mathematics Ⅰ，Linear Algebra and Analytical Geometry

学分：3

Credits：3

3.6.1 课程目的和基本内容（Course Objectives and Basic Contents）

本课程是储能科学与工程专业本科生的基础必修课，是一门综合性应用课程，是系统工程学科的重要理论和方法基础。它综合运用多种数学工具和定量化方法，结合实际问题

的特点与要求，对所研究的各类优化、决策问题建立合适的数学模型并设计有效的求解算法，其最终目标是为决策者提供定性理论工具和定量化方法工具以便于对实际问题做出最佳决策。

运筹学的主要研究内容包括对实际系统建立数学模型的方法和技术、有关数学模型的理论分析及求解算法设计与分析。本课程在规定的学时内，将主要介绍线性规划的单纯形法及对偶理论、整数规划、非线性规划、动态规划、图与网络优化等运筹学中的最基本内容。

This course is a basic compulsory course for undergraduates in energy storage science and engineering. As a comprehensive applied course, it covers an important theoretical and methodological foundation of systems engineering. It comprehensively uses a variety of mathematical tools and quantitative methods, combined with the characteristics and requirements of actual problems, to establish appropriate mathematical models and designs effective solving algorithms for various optimization and decision-making problems. Its goal is to provide decision makers with qualitative theoretical tools and quantitative method tools to make the best decision in applications.

The course focuses on methods and techniques for the formulation of mathematical models in actual systems, theoretical analysis of relevant mathematical models, and design and analysis of solving algorithms. This course, due to time limit, will mainly introduce the simplex algorithm and dual theory of linear programming, integer programming, non-linear programming, dynamic programming, graph and network optimization, etc.

3.6.2　课程基本情况（Basic Information of the Course ）

课程名称	运筹学 Operations Research										
开课时间	一年级		二年级		三年级		四年级		总学分	3	总学时	48
	秋	春	秋	春	秋	春	秋	春				
课程定位	储能科学与工程专业本科生数学基础课程群必修课											
授课学时分配	课堂讲授48学时											
先修课程	高等数学Ⅰ,线性代数与解析几何											
后续课程	信息物理融合能源系统,能源互联网											
教学方式	课堂教学,大作业,讨论,平时作业											
考核方式	课程结束笔试成绩占70%,大作业报告占20%,平时成绩占10%											
参考教材	[1] 刁在筠,刘桂真,戎晓霞,等. 运筹学. 4版. 北京:高等教育出版社,2016. [2] 希利尔,利伯曼.运筹学导论.9版. 北京:清华大学出版社,2010.											
参考资料	无											

3.6.3　教学目的和基本要求（Teaching Objectives and Basic Requirements）

（1）培养学生应用运筹学的思想方法分析和解决实际问题的意识与能力。学生应初步掌握对实际系统建立运筹学模型的方法和技术。

（2）培养学生对工程及科研中遇到的优化问题选择或设计有效算法的能力，培养学生根据数学模型结构、优化算法特性对优化问题的解进行解释的能力。学生应掌握线性规划、整数规划、非线性规划、动态规划、图与网络优化等运筹学中基本分支的基础理论、概念和相关算法。

（3）培养学生应用现代优化软件工具解决实际问题的能力。学生应掌握常见优化软件工具（LINGO、MATLAB 优化工具箱、CPLEX 等）的使用方法及特点，并能根据应用需求选择恰当的分析测试工具。

（1）Cultivate students' consciousness and ability to analyze and solve practical problems using operations research methods. Students should be able to build models using methods and techniques for model formulation from operations research for practical systems.

（2）Cultivate students' ability to select or design effective algorithms for optimization problems encountered in engineering and scientific research. Develop students' ability to explain solutions to optimization problems based on mathematical model structures and characteristics of optimization algorithms. Students should master the basic theories, concepts, and related algorithms of the basic branches of operations research such as linear programming, integer programming, nonlinear programming, dynamic programming, graph, and network optimization.

（3）Cultivate students' ability to use modern optimization software tools to solve practical problems. Students should master the usage methods and characteristics of common optimization software tools (LINGO, MATLAB optimization toolbox, CPLEX, etc.) and be able to choose appropriate analysis and testing tools according to application requirements.

3.6.4　课程大纲和知识点（Syllabus and Key Points）

第一章　绪论（introduction）

章节序号 chapter number	章节名称 chapters	知识点 key points	课时 class hour
1.1	概况 introduction	（1）运筹学的历史及特点 （2）主要分支、发展趋势 （1）history and characteristics of operations research （2）main branches and development trends	1

续表

章节序号 chapter number	章节名称 chapters	知识点 key points	课时 class hour
1.2	运筹学知识解决问题的基本原则 basic principles of operations research knowledge to solve problems	（1）建模与设计算法的基本原则 （2）模型正确性与有效性的概念 （3）应用运筹学解决问题的基本流程 （1）basic principles of modeling and algorithms design （2）concept of model's correctness and validity （3）basic flow of applying operations research to solve problems	1

第二章 线性规划（linear programming）

章节序号 chapter number	章节名称 chapters	知识点 key points	课时 class hour
2.1	线性规划问题及其标准形 linear programming and its standard form	（1）线性规划问题的实例 （2）线性规划问题的标准形 （1）examples of linear programming problems （2）standard form of linear programming problems	1
2.2	可行区域与基本可行解 feasible regions and basic feasible solutions	（1）二维问题的图解法 （2）可行区域的几何结构 （3）基本可行解 （1）graphical solution of two-variable linear programming problems （2）geometry of the feasible regions （3）basic feasible solutions	2
2.3	单纯形法 simplex method	（1）单纯形法的基本理论 （2）单纯形表 （1）basic theory of simplex method （2）simplex table	3
2.4	初始解 initial solution	（1）两阶段法 （2）大M法简介 （1）the two-phase method （2）the big M method	2
2.5	线性规划的对偶理论 duality theory of linear programming problems	（1）对偶线性规划的实例与定义 （2）对偶理论 （3）对偶单纯形法 （1）examples and definitions of dual linear programming problems （2）duality theory （3）dual simplex method	2

续表

章节序号 chapter number	章节名称 chapters	知识点 key points	课时 class hour
2.6	灵敏度分析 sensitivity analysis	（1）改变价值系数 （2）改变右端向量 （3）增加一个不等式约束 （1）change of the value coefficient （2）change of the right end vector （3）add an inequality constraint	1

第三章　整数规划（integer programming）

章节序号 chapter number	章节名称 chapters	知识点 key points	课时 class hour
3.1	整数规划简介 introduction to integer programming	（1）整数规划的基本建模方法 （2）基本概念 （3）整数规划问题求解的困难性 （4）整数规划问题求解的一般途径 （1）basic modeling methods for integer programming problems （2）basic concepts （3）difficulty in solving integer programming problems （4）general approaches to solving integer programming problems	2
3.2	Gomory割平面法 Gomory's cutting plane method	（1）Gomory割平面法的基本思想 （2）割平面法举例 （1）basic ideas underlying Gomory's cutting plane method （2）examples of cutting-plane method	1
3.3	分枝定界法 branch and bound algorithm	（1）分枝定界法的基本思想 （2）有关术语及算法步骤 （3）0-1规划的分枝定界法 （1）basic ideas underlying branch and bound algorithm （2）related terms and algorithm procedures （3）branch-and-bound method for 0-1 planning	2

第四章　非线性规划（nonlinear programming）

章节序号 chapter number	章节名称 chapters	知识点 key points	课时 class hour
4.1	基本概念 basic concepts	（1）非线性规划问题的实例 （2）基本概念 （3）下降类算法的一般结构 （1）examples of nonlinear programming problems （2）basic concepts （3）general structures of descending algorithms	2
4.2	凸函数与凸规划 convex functions and convex programming	（1）凸函数及其性质 （2）凸规划及其性质 （1）convex functions and their properties （2）convex programming and its properties	2
4.3	一维搜索 one-dimensional search	（1）0.618法 （2）牛顿法 （3）非精确一维搜索方法 （1）0.618 method （2）Newton method （3）inaccurate one-dimensional search method	3
4.4	无约束非线性规划 unconstrained nonlinear programming	（1）最优性条件 （2）最速下降法 （3）共轭梯度法 （4）拟牛顿法简介 （1）optimality condition （2）the steepest descent method （3）conjugate gradient method （4）Newton-like methods	4
4.5	约束非线性规划 constrained nonlinear programming	（1）光滑约束问题的最优性条件 （2）罚函数法 （3）可行方向法 （1）optimality conditions for smooth constraint problems （2）penalty function method （3）feasible direction method	4

第五章 动态规划（dynamic programming）

章节序号 chapter number	章节名称 chapters	知识点 key points	课时 class hour
5.1	最优性原理 principle of optimality	（1）多阶段决策问题的基本概念 （2）最优性原理及实例 （3）动态规划基本方程与最优性定理 （4）离散时间多阶段决策问题的动态规划方法总结 （1）basic concept of multi-stage decision-making problems （2）examples of optimal principles （3）basic equations and optimal principles of dynamic programming problems （4）summary of dynamic programming methods for discrete-time multi-stage decision-making problems	3
5.2	确定性定期多阶段决策问题 deterministic periodic multi-stage decision making problems	（1）投资分配问题 （2）TSP问题 （3）生产与库存控制问题 （4）工件排序问题 （1）investment allocation （2）traveling salesman problem (TSP) （3）production and inventory control （4）artifact sequencing problem	3
5.3	确定性不定期多阶段决策问题 deterministic unperiodic multi-stage decision-making problems	（1）函数空间迭代法 （2）策略空间迭代法 （1）function space iterative method （2）strategy space iterative method	1

第六章 网络规划（network programming）

章节序号 chapter number	章节名称 chapters	知识点 key points	课时 class hour
6.1	图与网络 graphs and networks	（1）图与网络的基本概念 （2）连通性与割集 （1）basic concepts of graphs and networks （2）connectivity and cut-sets	2
6.2	树与最小树 trees and minimal trees	（1）树与支撑树 （2）最小生成树及其求解算法 （1）trees and supporting trees （2）minimal spanning tree and its solving algorithm	2

续表

章节序号 chapter number	章节名称 chapters	知识点 key points	课时 class hour
6.3	最短有向路 the shortest directed path	（1）最短有向路 （2）迪杰斯特拉算法 （1）the shortest directed path （2）Dijkstra's algorithm	2
6.4	最大流 maximum flow	（1）最大流问题 （2）最大流算法 （1）maximum flow problem （2）maximum flow algorithm	2

_ 第 4 章 _

储能基础课程群

4.1 "大学物理(含实验)"教学大纲

课程名称：大学物理（含实验）
Course Title：Physics and Physics Experiments
先修课程：高等数学 I
Prerequisites：Advanced Mathematics I
学分：10
Credits：10

4.1.1 课程目的和基本内容（Course Objectives and Basic Contents）

大学物理（含实验）是储能科学与工程专业本科生的基础必修课。课程由两个部分组成，第一部分是大学物理，第二部分是大学物理实验。

第一部分是大学物理。物理学是研究物质的基本结构、相互作用和物质最基础、最普遍运动形式（机械运动、热运动、电磁运动、微观粒子运动等）及其相互转化规律的学科。物理学的研究对象具有极大普遍性，它的基本理论渗透在自然科学的一切领域、应用于生产技术的各个部门，它是自然科学许多领域和工程技术发展的基础。以物理学基础知识为内容的大学物理课程，它所包括的经典物理、近代物理和物理学在科学技术上应用的初步知识等都是一个高级工程技术人员必备的。因此，大学物理课是理工科各专业学生的一门重要必修基础课。

开设大学物理课程的目的，一方面在于为学生较系统地打好必要的物理基础；另一方面使学生初步学习科学的思想方法和研究问题的方法，这对开阔思路、激发探索和创新精神、增强适应能力、提高人才素质等，都会起到重要作用。学好物理，不仅对学生在校的学习十分重要，而且对学生毕业后的工作和进一步学习新理论、新技术，不断更新知识等，都将发挥深远影响。

第二部分是大学物理实验。大学物理实验是对理工科大学生进行科学实验基础训练的一门独立必修课程，包括力学、热学、电磁学、光学、原子物理等方面的基础实验内容，是一门实践性课程，是学生进入大学后接受系统实验方法和实验技能训练的开端。通过本课程的学习使学生了解实验的主要过程与基本方法，它旨在培养学生的科学素质、动手能力和创新能力等，为后续课程的学习和科研工作奠定必要的基础。本课程以基本物理量的测量方法、基本物理现象的观察和研究、常用测量仪器的结构和使用方法为主要内容进行教学，对学生的基本实验能力、分析能力、表达能力和综合性运用能力进行严格的培养。

Physics and Physics Experiments is a mandatory fundamental course for undergraduates majoring in energy storage science and engineering. It consists of two parts: 1. physics，2. physics experiments.

The first part is physics. Physics is a discipline that studies the basic structure, interaction, and the most basic and most universal motion forms of matter (mechanical motion, thermal motion, electromagnetic motion, microscopic particle motion, etc.) and their mutual transformation laws. The research objects of physics are extremely universal. Its basic theory permeates all fields of natural science and is applied to various departments of production technology. It is the basis for the development of many fields of natural science and engineering technology. The content of Physics is based on the basic knowledge of physics. The basic knowledge of classical physics, modern physics, and the application of physics in science and technology are all necessary for a senior engineer. Therefore, the physics course is an important mandatory basic course for all majors in science and engineering.

The purpose of the physics course is to lay a necessary physical foundation for students. On the other hand, this course enables students to initially learn scientific methods of thinking and researching. It also plays an important role to broaden ideas, stimulate exploration, innovation, and enhancing various adaptability. Learning physics is not only important for students' study at school, but also has a profound impact on students' work after graduation and further study of new theories, new technologies, and constantly updated knowledge.

The second part is the physics experiments, which is an independent mandatory course for the basic training for science and engineering students, including basic experimental content in mechanics, thermals, electromagnetics, optics, atomic physics, etc. It is a practical course for students to enter the university for receiving the beginning of systematic experimental methods and experimental skills training. Through the study of this course, students will understand the main processes and basic methods of scientific experiments. It aims to cultivate students' scientific quality, practical and innovative abilities, and lays the necessary foundation for students' follow-up course study and scientific research. The basic physical quantity measurement method, the observation and the research of the basic physical phenomena, the structure and method of the measuring instruments are the main contents for teaching, which can improve the basic experimental ability, analytical ability, expression ability, and comprehensive application ability for students.

4.1.2 课程基本情况（Basic Information of the Course）

课程名称	大学物理（含实验） Physics and Physics Experiments							
开课时间	一年级	二年级	三年级	四年级	总学分	10	总学时	192
	秋　春	秋　春	秋　春	秋　春				
课程定位	储能科学与工程专业本科生储能基础课程群必修课							
授课学时分配	课堂讲授128学时+实验64学时							
先修课程	高等数学Ⅰ							
后续课程	工程力学,自动控制理论,储能原理Ⅰ,储能热流基础（含实验）,传热传质学,流体机械原理及其储能应用,氢能储存与应用,半导体物理,电力系统分析							
教学方式	第一部分　大学物理:课堂教学、演示实验 第二部分　大学物理实验:教师讲解指导,学生独立完成实验							
考核方式	第一部分　大学物理:课程结束笔试成绩占50%,平时成绩占10%,第一次过程化考试占20%,第二次过程化考试占20% 第二部分　大学物理实验:每次实验课满分为20分,其中课内10分（包括预习、操作、数据及其他);实验报告10分（包括内容及格式、数据处理及结果、图表及曲线、分析及讨论和其他)。课程结束后,进行期末实验技能测试,测试成绩占总成绩的20%～40%							
参考教材	[1] 吴百诗. 大学物理学:上册. 北京:高等教育出版社,2014. [2] 吴百诗. 大学物理学:下册. 北京:高等教育出版社,2014. [3] 王红理,俞晓红,肖国宏. 大学物理实验. 西安:西安交通大学出版社,2018.							
参考资料	[1] 程守洙. 普通物理学. 北京:高等教育出版社,2016. [2] 张三慧. 大学物理学. 北京:清华大学出版社,2018.							

4.1.3 教学目的和基本要求（Teaching Objectives and Basic Requirements）

第一部分　大学物理

（1）为学生提供完整的基础物理学规律、知识体系,初步学习科学的思想方法和研究问题的方法,培养和训练学生以微积分为数学基础的发现问题、定量分析问题、解决科学问题的方法和能力;

（2）培养学生严肃的科学态度和求实的科学作风;

（3）培养学生终身学习的意识与能力,使学生具有跨越学科界限学习新知识、新方法的能力;

（4）为学生提供思想崇高、情怀宽广、责任感强烈的典型人物或事例,引导学生树立正确的价值观,培养具有爱国情怀和敢于担当的工程人才。

第二部分　大学物理实验

（1）通过本课程的学习,使学生接受系统的实验方法和实验技能训练,锻炼学生的实验动手能力，提高科学素质，并培养严谨的治学态度、活跃的创新意识和理论联系实际的动手能力，为后续课程学习和科研工作奠定必要的基础；

（2）要求学生熟悉实验课程的完整环节，掌握基本物理量的测量方法与手段，以及现代常用实验仪器的原理、调节和使用，掌握实验误差和不确定度计算的基本知识，具有正确处理实验数据的能力，能够撰写合格的实验报告，独立、自主地进行基础、综合与设计性实验。

The first part: physics

（1）Provide students with a complete basic physics laws and knowledge system, a preliminary study of scientific thinking methods and methods of researching problems, cultivate and train students' ability to discover problems, analyze problems quantitatively and solve scientific problems based on calculus.

（2）Cultivate students' serious scientific attitude and realistic scientific style.

（3）Cultivate students' awareness and ability of lifelong learning, so that students can learn new knowledge and new methods across subject boundaries.

（4）Provide students with typical characters or examples with lofty thinking, broad sentiment, and strong sense of responsibility. Guide students to establish correct values, and cultivate engineering talents with patriotism and courage.

The second part：physics experiments

（1）Through the study of this course, students will receive systematic training of experimental methods and experimental skills. This will exercise students' experimental ability, improve their scientific quality, and cultivate a rigorous academic attitude, active innovation consciousness, and the ability to integrate theory with practice. Lay the necessary foundation for follow-up courses study and scientific research work.

（2）Students are required to be familiar with the complete links of experimental courses, master basic physical quantity measurement methods, their principle, adjustment and use of modern experimental instruments. Students are expected to master the basic knowledge of experimental error and uncertainty analysis, and can correctly process experimental data. Be able to write qualified experiment reports, design and conduct basic and comprehensive experiments independently.

4.1.4 课程大纲和知识点（Syllabus and Key Points）

第一章 力学（mechanics）

章节序号 chapter number	章节名称 chapters	知识点 key points	课时 class hour
1.1	质点运动学 particle kinematics	（1）质点、刚体等模型和参照系、惯性系等概念 （2）位置矢量、位移、速度、加速度等物理量 （3）牛顿力学的相对性原理,相对运动和伽利略变换 （1）models such as mass point and rigid bodies, reference systems, and inertial systems, etc. （2）physical quantities such as position vector, displacement, velocity, acceleration, etc. （3）relativity principle of Newtonian mechanics, the relative motion and Galilean transformation	5
1.2	质点动力学 particle dynamics	（1）牛顿运动定律及其适用条件,用牛顿第二定律的投影式处理力学问题 （2）功的概念,变力的功,力矩的功,保守力作功的特点及势能的概念,势能和势能差,重力势能、弹性势能和万有引力势能 （3）质点系的动能定理和动量定理,质心和质心运动定理,空间均匀性与动量守恒律的关系 （4）机械能守恒定律、动量守恒定律及它们的适用条件 （1）Newton's law of motion and its applicable conditions, using the projection law of Newton's second law to deal with mechanical problems （2）concept of work, variable force work and torque, the characteristics of conservative work and the concept of potential energy, potential energy and potential energy difference, and skillfully gravitational potential energy, elastic potential energy, and universal gravitational potential energy （3）kinetic energy theorem and the momentum theorem of the particle system, centroid and centroid motion theorem, relationship between spatial uniformity and momentum conservation law （4）law of conservation of mechanical energy, law of conservation of momentum, and their applicable conditions	12

续表

章节序号 chapter number	章节名称 chapters	知识点 key points	课时 class hour
1.3	刚体的运动 rigid body movement	（1）转动惯量概念,能用积分方法计算几何形状简单、质量分布均匀的物体对轴的转动惯量,力矩和刚体绕定轴的转动定律 （2）会计算定轴转动刚体的动能,能运用定轴转动刚体的动能定理分析、计算简单力学问题 （3）动量矩（角动量）概念;动量矩守恒定律及其适用条件;进动产生的原因和进动方向 （1）concept of moment of inertia. The integral method can be used to calculate the moment of inertia of an object with simple geometry and a uniform mass distribution. The moment of force and the law of rotation of a rigid body around a fixed axis （2）calculate the kinetic energy of a rigid body rotating on a fixed axis, and use the kinetic energy theorem of a rigid body rotating on a fixed axis to analyze and calculate simple mechanical problems （3）concept of momentum moment (angular momentum), law of conservation of momentum and its applicable condition, cause of precession and precession direction	9

第二章　气体动理论及热力学（kinetic theory of gas and thermodynamics）

章节序号 chapter number	章节名称 chapters	知识点 key points	课时 class hour
2.1	热力学 thermodynamics	（1）功、热量和内能的概念,准静态过程,热力学第一定律 （2）可逆过程与不可逆过程,热力学第二定律的两种表述 （1）concepts of work, heat, and internal energy, quasi-static process, the first law of thermodynamics （2）reversible process and the irreversible process, two expressions of the second law of thermodynamics	10
2.2	气体动理论 kinetic theory of gas	（1）玻尔兹曼能量分布定律,麦克斯韦速率分布定律、速率分布函数及速率分布曲线的物理意义,气体分子热运动的算术平均速率、均方根速率和最概然速率的求法和意义,重力场中粒子按高度的分布规律	10

续表

章节序号 chapter number	章节名称 chapters	知识点 key points	课时 class hour
2.2	气体动理论 kinetic theory of gas	（2）理解气体分子能量按自由度均分定理 （3）阿伏伽德罗常数、玻尔兹曼常数的数值和单位,常温常压下气体分子数密度、平均自由程和平均碰撞频率 （1）Boltzmann's law of energy distribution, Maxwell's law of rate distribution, rate distribution function and rate distribution curve; the arithmetic average velocity, root mean square velocity, and most probable velocity of the thermal motion of gas molecules, distribution law of particles in the field according to height in the gravity （2）degree of freedom and the equipartition theorem （3）values and units of the Avogadro constant and the Boltzmann constant, order of magnitude of gas molecular number density, arithmetic mean rate, mean free path and mean collision frequency	10

第三章　电磁学（electromagnetism）

章节序号 chapter number	章节名称 chapters	知识点 key points	课时 class hour
3.1	静电场 electrostatic field	（1）静电场电场强度和电势的概念以及场的叠加原理 （2）静电场的规律:高斯定理和环路定理 （3）静电平衡条件及导体处于静电平衡时的基本特征 （1）concept of electric field intensity and potential of the electrostatic field and the superposition principle of the field （2）law of electrostatic field: Gauss theorem and loop theorem （3）electrostatic balance conditions and the basic characteristics of the conductor in electrostatic equilibrium	10
3.2	稳恒电流的磁场 magnetic field induced by steady current	（1）磁感应强度的概念及毕奥-萨伐尔定律 （2）安培定律和洛伦兹力公式,电偶极矩和磁矩的概念 （3）介质的极化、磁化现象及其微观解释,介质中的高斯定理和安培环路定理 （1）concept of magnetic intensity and Biot-Savart law （2）Ampere's law and Lorentz force formula, concepts of electric dipole moment and magnetic moment （3）polarization, magnetization, and microscopic interpretation of the medium, Gauss theorem and Ampere's loop theorem in the medium	10

续表

章节序号 chapter number	章节名称 chapters	知识点 key points	课时 class hour
3.3	电磁感应 electromagnetic induction	（1）电动势的概念,法拉第电磁感应定律,动生电动势和感生电动势的概念和规律 （2）电容、自感系数和互感系数的定义及其物理意义 （1）concept of electromotive force, Faraday law of electromagnetic induction, concepts and laws of motional electromotive force and induced electromotive force （2）definition and physical meaning of capacitance, self-inductance coefficient, and mutual inductance coefficient	6
3.4	电磁场理论的基本概念 basic concept of electromagnetic field theory	（1）电磁场的物质性,电能密度、磁能密度的概念 （2）有旋电场和位移电流,对称分布场中有旋电场的场强和位移电流,麦克斯韦积分方程的物理意义 （1）materiality of electromagnetic fields, concepts of electrical energy density and magnetic energy density （2）rotating electric field and displacement current, field strength and displacement current of rotating electric field in symmetric distribution field, physical meaning of Maxwell's integral equations	4

第四章　机械振动和机械波（mechanical vibration and mechanical waves）

章节序号 chapter number	章节名称 chapters	知识点 key points	课时 class hour
4.1	机械振动 mechanical vibration	（1）谐振动和简谐波的物理量 （2）旋转矢量法 （3）两个同方向同频率谐振动的合成规律,两个同方向不同频率谐振动的合成规律 （1）physical quantities of harmonic vibration and harmonics （2）rotation vector method （3）synthesis law of two harmonic vibrations in the same direction and frequency, synthesis law of two harmonic vibrations with different frequencies in the same direction	6

续表

章节序号 chapter number	章节名称 chapters	知识点 key points	课时 class hour
4.2	机械波 mechanical wave	（1）惠更斯原理和波的叠加原理,波的相干条件 （2）驻波的概念及其形成条件,驻波和行波的区别,驻波的相位分布特点 （3）机械波的多普勒效应 （1）Huygens' principle, superposition principle and coherence condition of waves （2）concept of standing wave and its forming conditions, difference between standing wave and traveling wave, phase distribution characteristics of standing wave （3）Doppler effect of mechanical waves	8

第五章　波动光学（wave optics）

章节序号 chapter number	章节名称 chapters	知识点 key points	课时 class hour
5.1	光的干涉 interference of light	（1）普通光源的发光机理,获得相干光的方法 （2）光程概念以及光程差和相位差的关系,半波损失问题,迈克耳孙干涉仪的原理 （1）luminescence mechanism of ordinary light sources and the method of obtaining coherent light （2）concept of optical path and the relationship between optical path difference and phase difference, half wave loss, principle of the Michelson interferometer	12
5.2	光的衍射 diffraction of light	（1）惠更斯-菲涅耳原理,单缝衍射条纹的亮度分布规律 （2）圆孔的夫琅禾费衍射 （3）光栅衍射公式 （1）Huygens-Fresnel's principle, diffraction of single slit （2）Fraunhofer diffraction of a circular hole （3）diffraction of grating	
5.3	光的偏振 polarization of light	（1）自然光和线偏振光,偏振光的获得方法和检验方法,马吕斯定律和布儒斯特定律 （2）双折射现象及偏振光的干涉 （1）natural light and linearly polarized light, obtaining methods and test methods of polarized light, Malus' law and Brewster's law （2）phenomenon of birefringence and interference of polarized light	4

第六章　近代物理（modern physics）

章节序号 chapter number	章节名称 chapters	知识点 key points	课时 class hour
6.1	狭义相对论力学基础 introduction to special relativistic mechanics	（1）爱因斯坦狭义相对论的两个基本假设,洛伦兹坐标变换 （2）狭义相对论同时性的概念、长度收缩概念和时间膨胀概念,牛顿力学时空观和狭义相对论时空观以及二者的差异 （3）理论狭义相对论中质量和速度的关系、质量和能量的关系 （1）two basic assumptions of Einstein's special theory of relativity and Lorentz coordinate transformation （2）concepts of simultaneity of relativity, length contraction, and time expansion, space-time view of Newtonian mechanics and the space-time view of special relativity and their difference （3）relationship between mass and velocity, relationship between mass and energy in the theory of special relativity	8
6.2	量子物理基础 introduction to quantum physics	（1）普朗克的能量子假设,光电效应和康普顿效应的实验规律以及爱因斯坦的光子理论对这两个效应的解释,光的波粒二象性 （2）氢原子光谱的实验规律,波尔氢原子理论,波尔氢原子理论的意义和局限性 （3）德布罗意物质波假设和电子衍射实验,实物粒子的二象性,物质波动性物理量和粒子性物理量之间的关系 （4）波函数及其统计解释,一维坐标动量不确定关系 （5）能量量子化、角动量量子化及空间量子化,施特恩—革拉赫实验及微观粒子的自旋,描述原子中电子运动状态的四个量子数,泡利不相容原理和原子的电子壳层结构 （6）一维定态薛定谔方程以及在一维无限深势井情况下薛定谔方程的解 （1）Planck's energy quanta hypothesis, experimental laws of the photoelectric effect and the Compton effect, and the explanation of these two effects by Einstein's photon theory, wave-particle duality of light	10

章节序号 chapter number	章节名称 chapters	知识点 key points	课时 class hour
6.2	量子物理基础 introduction to quantum physics	（2）experimental laws of hydrogen atom spectrum, Bohr's hydrogen atom theory, significance and limitations of Bohr's hydrogen atom theory （3）De Broglie's matter-wave hypothesis and electron diffraction experiment, wave-particle duality of real particles, discribe the relationship between the fluctuating physical quantity of matter and the particle physical quantity （4）wave function and its statistical interpretation, uncertainty principle （5）energy quantization, angular momentum quantization, and spatial quantization, Stern-Gralach experiments and the spins of microscopic particles, the four quantum numbers describing the state of electron motion in an atom, and Pauli incompatibility principle and atomic electronic shell structure （6）one-dimensional stationary Schrödinger equation, solution of the Schrödinger equation in the case of one-dimensional infinite potential wells	10
6.3	固体、激光、核物理与粒子物理简介 introduction to solid, laser, nuclear physics and particle physics	（1）固体能带的形成,本征半导体、n型半导体和p型半导体 （2）激光的形成、特性及其主要应用 （3）原子核的构造和性质,核力与结合能,基本粒子的组成和特性等 （1）formation of energy bands in solid, intrinsic semiconductors, n-type semiconductors, and p-type semiconductors （2）formation, characteristics, and main applications of lasers （3）structure and properties of the nucleus, nucleus forces and binding energies, composition and characteristics of elementary particles, etc.	4

4.1.5　实验环节（Experiments）

序号 number	实验内容 experiment content	知识点 key points	课时 class hour
1	误差与不确定度理论及物理实验基本知识、力学与热学实验 error and uncertainty theory, and basic knowledge of physics experiments, mechanics and thermal experiments	（1）实验导论 （2）力学基本测量系列实验 （3）力学振动与波系列实验 （4）热学流体系列实验 （5）热学温度场系列实验 （1）introduction of the experiment （2）mechanical basic measurement series experiments （3）mechanical vibration and wave series experiments （4）thermal and fluid series experiments （5）thermal temperature field series experiments	16
2	电磁学实验 electromagnetic experiment	（1）电磁学示波器系列实验 （2）电磁学元器件特性系列实验 （3）电磁学电桥系列实验 （4）电磁学电磁场系列实验 （1）electromagnetic and oscilloscope series experiments （2）electromagnetic component characteristics series experiments （3）electromagnetic bridge series experiments （4）electromagnetic field series experiments	16
3	声学及光学实验 acoustics and optical experiments	（1）声学超声系列实验 （2）光学调节系列实验 （3）光学综合性系列实验 （4）光学分光计系列实验 （1）acoustic and ultrasound series experiments （2）optical adjustment series experiments （3）optical comprehensive series experiments （4）optical spectrometer series experiments	16
4	近代与综合实验 modern and comprehensive experiments	（1）综合性光电系列实验 （2）综合性功能材料系列实验 （3）综合性位移传感器系列实验 （4）虚拟仿真系列实验 （1）comprehensive photoelectric series experiments （2）comprehensive functional material series experiments （3）comprehensive displacement sensor series experiments （4）virtual simulation series experiments	16

4.2 "工程力学"教学大纲

课程名称：工程力学

Course Title：Engineering Mechanics

先修课程：高等数学Ⅰ，大学物理（含实验）

Prerequisites：Advanced Mathematics Ⅰ，Physics and Physics Experiments

学分：3

Credits：3

4.2.1 课程目的和基本内容（Course Objectives and Basic Contents）

本课程是高等学校工科各专业的一门重要的技术基础课，与工程实际有着非常密切的关系，是解决工程实际问题的基础，也是一系列后续课程的基础，在诸多工程技术领域中有着广泛的应用。

课程内容包括静力学，杆件基本变形的强度、刚度及稳定性计算和运动学三部分内容。

This course is an important technical fundamental course for all majors of engineering in colleges and universities, closely related to the engineering practice. It is the basis of solving practical engineering problems and a series of follow-up courses，which is widely used in many engineering and technical fields.

The course focuses on statics, calculation of basic deformation strength, stiffness and stability of bars, and kinematics.

4.2.2 课程基本情况（Basic Information of the Course）

课程名称	工程力学 Engineering Mechanics											
开课时间	一年级		二年级		三年级		四年级		总学分	3	总学时	48
	秋	春	秋	春	秋	春	秋	春				
课程定位	储能科学与工程专业本科生储能基础课程群必修课											
授课学时分配	课堂讲授48学时+课外20学时（课外学时不计入总学时）											
先修课程	高等数学Ⅰ，大学物理（含实验）											
后续课程	工程图学,半导体物理,储能原理II,流体机械原理及其储能应用											
教学方式	课堂教学,课内外讨论,课外自学,团队合作											
考核方式	课程结束笔试成绩占70%,平时成绩占30%											
参考教材	[1]冯立富,伍晓红,刘百来,等.工程力学.2版.西安:西安交通大学出版社,2020.											
参考资料	[1]奚绍中,邱秉权,沈火明.工程力学教程.4版.北京:高等教育出版社,2019.											

4.2.3　教学目的和基本要求（Teaching Objectives and Basic Requirements）

（1）掌握刚体静力学、点和刚体运动学的基本知识和理论，建立固体力学的基本概念（包括位移、变形、应变、内力、应力、弹性、塑性、承载能力等），掌握杆件基本变形的强度、刚度及稳定性的计算。

（2）初步学会应用工程力学的理论和方法分析、解决一些简单工程实际问题，培养学生的工程意识、对工程实际问题的力学建模能力及分析计算能力。

（3）培养学生实验研究能力、团队合作能力、科学的思维方式及正确的世界观。

（1）Master the basic knowledge and theory of rigid body statics, particle and rigid body kinematics. Establish basic concepts of solid mechanics (including displacement, deformation, strain, internal force, stress, elasticity, plasticity, bearing capacity, etc.). Master the calculation of basic deformation strength, stiffness, and stability of bars.

（2）Learn how to apply the theory and method of engineering mechanics to analyze and solve some simple engineering problems. Develop engineering consciousness, the ability of mechanical modeling for solving and analyzing practical engineering problems.

（3）Cultivate the ability of experimental research, teamwork, scientific thinking, and correct world outlook.

4.2.4　课程大纲和知识点（Syllabus and Key Points）

第一章　静力学基础（fundamentals of statics）

章节序号 chapter number	章节名称 chapters	知识点 key points	课时 class hour
1.1	绪论 introduction	（1）力 （2）力的表示法 （1）force （2）representation of force	1
1.2	静力学基本概念 basic concept of statics	（1）刚体 （2）变形体 （3）力系平衡公理 （1）rigid body （2）deformed body （3）equilibrium axiom of force system	1

续表

章节序号 chapter number	章节名称 chapters	知识点 key points	课时 class hour
1.3	约束和约束力 constraint and constrained force	（1）约束 （2）约束力 （1）constraint （2）constraint force	1
1.4	受力分析 force analysis	（1）受力分析 （2）受力分析图 （1）force analysis （2）force analysis diagram	1

第二章　平面基本力系（fundamental coplanar force system）

章节序号 chapter number	章节名称 chapters	知识点 key points	课时 class hour
2.1	基本力系的合成与平衡 equilibrium and composition of fundamental force system	（1）基本力系的平衡 （2）基本力系的合成 （1）equilibrium of fundamental force system （2）composition of fundamental force system	1

第三章　平面任意力系（arbitrary coplanar force system）

章节序号 chapter number	章节名称 chapters	知识点 key points	课时 class hour
3.1	平面力系的简化与平衡 equilibrium and simplification of coplanar force system	（1）平面力系的简化 （2）平面力系的平衡 （1）simplification of coplanar force system （2）equilibrium of coplanar force system	1
3.2	物体系的平衡 equilibrium of object system	（1）物体系的平衡 （2）静定问题概念 （3）静不定问题概念 （1）equilibrium of object system （2）concepts of statically determinate （3）concepts of statically indeterminate	2
3.3	平面桁架内力分析 analysis of internal forces of plane truss	平面桁架内力分析 analysis of internal forces of plane truss	1

第四章 摩擦（friction）

章节序号 chapter number	章节名称 chapters	知识点 key points	课时 class hour
4.1	摩擦 friction	（1）摩擦 （2）考虑摩擦的平衡问题 （3）滚动摩阻 （1）friction （2）equilibrium problem considering friction （3）rolling friction	2

第五章 空间力系（spatial force system）

章节序号 chapter number	章节名称 chapters	知识点 key points	课时 class hour
5.1	空间力系的简化与平衡 equilibrium and simplification of spatial force system	（1）空间力系的简化与平衡 （2）主矢、主矩、力矩、力偶矩 （1）equilibrium and simplification of spatial force system （2）main vector, main moment of force, moment of force, moment of force couple	1
5.2	重心和形心 barycenter and centroid	（1）重心 （2）形心 （1）barycenter （2）centroid	1

第六章 拉伸与压缩（tension and compression）

章节序号 chapter number	章节名称 chapters	知识点 key points	课时 class hour
6.1	变形体的基本概念与假设 basic concept and hypothesis of deformable body	（1）变形体的基本概念 （2）变形体假设 （1）basic concept of deformable body （2）hypothesis of deformable body	1
6.2	拉压内力、应力 internal force and stress of tension and compression	（1）轴力图 （2）横截面和斜截面上的应力 （1）axial force diagram （2）stress on cross-section and inclined section	2

续表

章节序号 chapter number	章节名称 chapters	知识点 key points	课时 class hour
6.3	拉压强度、变形计算 calculation of tensile and compressive strength and deformation	（1）拉压的强度计算、变形计算 （2）杆内的应变能 （1）calculation of tensile and compressive strength and deformation （2）strain energy in bars	2
6.4	材料力学性能 mechanical properties of materials	（1）材料受压力学性能 （2）材料受拉力学性能 （1）compression mechanical properties of materials （2）tensile mechanical properties of materials	1
6.5	拉压超静定 statically indeterminate of tension and compression	拉压超静定问题 statically indeterminate of tension and compression	1
6.6	拉压杆接头计算 calculation of tensile compressive bar joint	（1）圣维南原理 （2）拉压杆接头计算 （1）Saint-Venant's principle （2）calculation of tensile compressive bar joint	1

第七章　扭转（torsion）

章节序号 chapter number	章节名称 chapters	知识点 key points	课时 class hour
7.1	扭转内力应力计算 calculation of internal force and stress of torsion	（1）扭矩图 （2）圆轴横截面切应力 （1）torque diagram （2）shear stress of cross-section of circular shaft	1
7.2	扭转强度计算 calculation of torsional strength	（1）圆轴扭转强度和刚度条件 （2）圆轴扭转破坏分析 （1）torsional strength and stiffness conditions of circular shaft （2）torsional failure analysis of circular shaft	1

第八章 弯曲（bending）

章节序号 chapter number	章节名称 chapters	知识点 key points		课时 class hour
8.1	截面图形的几何性质 geometric properties of section figure	静矩和惯性矩 static moment and moment of inertia		1
8.2	弯曲内力计算 calculation of bending internal force	弯曲内力图 bending internal force diagram		2
8.3	弯曲正应力 bending normal stress	（1）弯曲正应力 （2）强度校核 （1）bending normal stress （2）strength check		2
8.4	弯曲变形 bending	（1）梁的强度条件 （2）弯曲变形 （1）strength condition of beam （2）bending		1
8.5	弯曲应变能、超静定梁 bending strain energy, statically indeterminate beam	（1）弯曲应变能 （2）超静定梁 （1）bending strain energy （2）statically indeterminate beam		2

第九章 压杆稳定（stability of members）

章节序号 chapter number	章节名称 chapters	知识点 key points		课时 class hour
9.1	压杆临界力、临界应力 critical force and critical stress of compression bars	（1）稳定性 （2）细长压杆的临界载荷 （1）stability （2）critical load of slender compression bars		2
9.2	稳定性校核 stability check	稳定性校核 stability check		1

第十章 点的运动学（kinematics of points）

章节序号 chapter number	章节名称 chapters	知识点 key points	课时 class hour
10.1	点的运动学 kinematics of points	（1）描述点的运动的矢量法、直角坐标法和自然坐标法 （2）点的运动轨迹 （3）点的速度、加速度 （1）vector method, rectangular coordinate method, and natural coordinate method used to describe the motion of a point （2）trajectory of a point （3）velocity and acceleration of a point	1

第十一章 刚体基本运动（fundamental motion of a rigid body）

章节序号 chapter number	章节名称 chapters	知识点 key points	课时 class hour
11.1	刚体基本运动 fundamental motion of a rigid body	（1）刚体移动 （2）刚体定轴转动 （1）motion of a rigid body （2）fixed axis rotation of a rigid body	1

第十二章 点的复合运动（composite motion of points）

章节序号 chapter number	章节名称 chapters	知识点 key points	课时 class hour
12.1	复合运动的基本概念 basic concept of composite motion	（1）运动合成的基本概念 （2）运动分解的基本概念 （1）concept of motion composition （2）concept of motion decomposition	2
12.2	速度合成定理 theorem for composition of velocities	速度合成定理 theorem for composition of velocities	2
12.3	加速度合成定理 theorem of composition of acceleration	加速度合成定理 theorem for composition of acceleration	2

第十三章　刚体平面运动（plane motion of a rigid body）

章节序号 chapter number	章节名称 chapters	知识点 key points	课时 class hour
13.1	刚体平面运动的基本概念 basic concept of plane motion of a rigid body	（1）刚体平面运动方程 （2）刚体平面运动分解 （1）plane motion equation of rigid body （2）motion decomposition of rigid body	2
13.2	速度分析 velocity analysis	刚体平面运动的速度分析 velocity analysis of plane motion of rigid body	2
13.3	加速度分析 acceleration analysis	刚体平面运动的加速度分析 acceleration analysis of plane motion of rigid body	2

4.3　"储能化学基础（含实验）"教学大纲

课程名称：储能化学基础（含实验）

Course Title：General Chemistry for Energy Storage（with Experiments）

先修课程：无

Prerequisites：None

学分：7.5

Credits：7.5

4.3.1　课程目的和基本内容（Course Objectives and Basic Contents）

第一部分　储能化学基础

本课程为储能科学与工程专业提供坚实的化学基础。立足对物质本质的认识，深入介绍原子、分子近代结构理论，讨论分子之间的作用力及结构与基本性质的关系，归纳、提炼出自然科学中最基础、最本质的原理及变化规律，使学生对自然科学有一个更加深入的认知。

课程主要内容包括物质结构、溶液与胶体、热力学第一定律、热力学第二定律、化学平衡与反应动力学、表面现象、离子型反应、自由基反应和聚合反应、氧化还原反应、能源化学基础、含能物质基础、储能化学相关材料简介。

第二部分　储能化学基础实验

通过对储能化学基础实验的学习，使学生了解并掌握与专业相关的无机物、有机物、高分子材料等的制备方法，了解常见的物质分析和表征方法。培养学生实事求是的科学态度和良好的科学素养，为进一步从事储能及相关领域的学习和研究奠定基础。培养学生理论联系实际的能力、终身学习意识和自我管理与学习能力。

课程主要内容包括实验室安全、无机物、溶胶制备、酸碱滴定、熔沸点、蒸汽压测定，以及环己烯、溴乙烷、乙酰水杨酸的制备。

The first part is general chemistry for energy storage.

The course aims to provide a basic knowledge of chemistry for students majoring in energy storage science and engineering. It introduces the advanced structural theory of atoms and molecules, and discusses the intermolecular forces and relationship between structure and properties. Essential principles and change rules in natural science could be summarized and refined, based on the understanding of the nature of matter. It enables students to have a deeper understanding of natural science.

The main contents of the course include matter structure, solution and colloid, the first and second laws of thermodynamics, chemical equilibrium and reaction kinetics, surface phenomena, ionic reactions, free radical reactions and polymerization reactions, redox reaction, fundamentals of energy chemistry, fundamentals of energetic material, and introduction of energy storage chemistry related materials.

The second part is experiments of general chemistry for energy storage.

Through learning of experiments of general chemistry for energy storage, students can master the preparation skills of inorganics, organics, as well as polymer materials, and know the common method for analysis and characterization of materials. This course aims to cultivate the scientific attitude of "seek truth from facts", and good scientific literacy, which establishes a foundation for further study and research in energy storage. It trains students' ability to link theory with practice, life-study consciousness, and self-management and learning ability.

The main contents of the course include laboratory safety,inorganics, sol preparation, acid-base titration, determination of melting/boiling point and vapor pressure; preparations of cyclohexene, bromoethane, and acetylsalicylic acid.

4.3.2 课程基本情况（Basic Information of the Course）

课程名称	储能化学基础（含实验） General Chemistry for Energy Storage（with Experiments）											
开课时间	一年级		二年级		三年级		四年级		总学分	7.5	总学时	144
	秋	春	秋	春	秋	春	秋	春				
课程定位	储能科学与工程专业本科生储能基础课程群必修课											
授课学时分配	课堂讲授96学时+实验48学时											
先修课程	无											

续表

后续课程	储能原理 I , 氢能储存与应用, 电化学基础, 电池材料制备技术, 储能电池设计、制作及集成化实验
教学方式	课堂教学、实验与综述报告、讨论
考核方式	储能化学基础 平时:40%（平时测验成绩占30%,平时作业占10%） 期末:60% 储能化学基础实验 平时:30%（平时测验成绩占10%,平时作业占20%） 期末:70%
参考教材	[1] 张志成,张雯. 大学化学. 北京:科学出版社,2018. [2] 徐寿昌. 有机化学. 北京:高等教育出版社,2014. [3] 王明德. 物理化学.2版. 北京:化学工业出版社,2015. [4] 郑阿群,孙杨. 大学化学实验 .西安:西安交通大学出版社,2018. [5] 郗英欣,白艳红. 有机化学实验. 西安:西安交通大学出版社,2014. [6] 王明德,王耿,吴勇. 物理化学实验. 西安:西安交通大学出版社,2013.
参考资料	[1] Silberberg M S. Principles of General Chemistry.2nd ed.New York: McGraw-Hill Companies,Inc., 2008. [2] 天津大学物理化学教研室. 物理化学.6版.北京:高等教育出版社,2017. [3] 潘祖仁. 高分子化学.5版.北京:化学工业出版社,2014. [4] 庄继华,等. 物理化学实验.3版. 北京:高等教育出版社,2004. [5] 梁晖,卢江. 高分子化学实验. 北京:化学工业出版社,2016.

4.3.3 教学目的和基本要求（Teaching Objectives and Basic Requirements）

第一部分 储能化学基础

（1）通过课程学习，使学生熟悉近代原子结构模型、价键理论、分子轨道理论等物质结构的基本理论，理解化学反应中物质组成、结构和性质的关系，并能应用这些理论解释物质的结构与性质的关系及其变化规律。熟悉气体、液体、溶液和胶体的性质。

（2）掌握化学热力学的基本概念以及在化学反应的能量转化、化学反应的方向和限度问题上的应用；掌握化学反应的速率和反应机理；能够从宏观和微观的不同角度理解化学变化的基本特征，培养逻辑思维能力和批判性思维精神。

（3）熟悉常见有机化合物的结构和性质，掌握链烃（烷、烯、炔）、芳香烃的结构及化学性质；掌握离子型反应（沉淀反应、酸碱反应、配位反应、亲电反应和亲核反应）、自由基反应、聚合反应的基本知识和基本原理。

（4）了解能源的分配和相互转化、含能物质的分类和合成方法。了解储氢材料、电催化材料的分类和基本特征。了解锂电池、电容器的工作原理和电极材料。

（5）培养学生自学和归纳总结的能力、运用所学知识分析和解决问题的能力及团队合作能力。

（6）培养学生理论联系实际的能力和实事求是的科学态度，终身学习意识和自我管理、自主学习能力。

第二部分　储能化学基础实验

（1）通过对储能化学基础实验的学习，使学生掌握与专业相关的无机物、有机物、高分子材料等的制备方法，了解常见的物质分析和表征方法。

（2）掌握不同物质制备的基本原理和基本技巧，了解专业涉及的物质合成的前沿发展动态。

（3）通过实验促进学生对理论知识的理解及运用，培养学习者自学和归纳总结的能力及运用所学分析问题和解决问题的能力，培养团队合作能力。

（4）培养学生实事求是的科学态度和良好的科学素质，为进一步从事储能及相关领域的学习和研究奠定基础，培养学生理论联系实际的能力和实事求是的科学态度，终身学习意识和自我管理、自主学习能力。

The first: general chemistry for energy storage

（1）Through the study of general chemistry for energy storage, students should be familiar with the basic theories of matter structure such as modern atomic structure models, valence-bond theory, molecular orbital theory, etc., and understand the relationship between the composition, structure and properties of materials in chemical reactions. To apply these theories to explain the relationship between the structure and the properties of matter and their changing rules. Be familiar with the properties of gas, liquid, solution and colloid.

（2）Grasp the basic concepts of chemical thermodynamics and their applications in the chemical reaction (such as energy conversion, direction and limit). Master the rate mechanism and dynamical problems of chemical reactions. Understand the basic characteristics of chemical change from different perspectives of macro and micro, and cultivate logical thinking ability and critical thinking spirit.

（3）Be familiar with the structure and properties of common organic compounds, master the structure and chemical properties of chain hydrocarbons (alkanes, alkenes, alkynes), aromatic hydrocarbons; master the basic knowledge and basic principles of ionic reactions (precipitation reactions, acid-base reactions, coordination reactions, electrophilic reactions and nucleophilic reactions), free radical reactions and polymerization reactions.

（4）Understand the distribution and mutual conversion of energy, the classification and synthesis methods of energy sources. Understand the classification and basic characteristics of hydrogen storage materials and electro catalytic materials. Understand the working principles and electrode materials of lithium batteries and capacitors.

（5）Develop the ability of self-study and summary, the ability to analyze and solve problems, and the ability of teamwork.

（6）Train the ability to connect theory with practice. Cultivate students' practical scientific attitude, life-study consciousness, and the ability of self-management and independent learning.

The second part: experiments of general chemistry for energy storage

（1）Master the knowledge on the preparation methods of inorganics, organics, and polymer materials, and understand the analysis and characterization methods for common materials.

（2）Master the basic principles and skills of the preparation of substances, and understand the frontier development of substance synthesis.

（3）Through experiments, promote the understanding and application of theoretical knowledge, and strengthen the ability of self-study, induction, summarization, analysis and problem-solving ability, and teamwork.

（4）Develop scientific attitude of "seek truth from facts", and good scientific literacy. Establish a foundation for further study and research in energy storage. Train the ability to link theory with practice, life-study consciousness, self-management, and learning ability.

4.3.4 课程大纲和知识点（Syllabus and Key Points）

绪论（introduction）

章节序号 chapter number	章节名称 chapters	知识点 key points	课时 class hour
0	绪论 introduction	（1）课程规划 （2）化学的特征和地位 （1）course planning （2）characteristics and status of chemistry	1

第一章 原子结构（atomic structure）

章节序号 chapter number	章节名称 chapters	知识点 key points	课时 class hour
1.1	近代原子结构发展历程 the development history of modern atomic structure	（1）原子能级 （2）波粒二象性 （3）原子轨道（波函数）和电子云 （1）atomic energy level （2）wave-particle duality （3）atomic orbit (wave function) and electron cloud	1

续表

章节序号 chapter number	章节名称 chapters	知识点 key points	课时 class hour
1.2	现代原子结构模型 modern atomic structure model	（1）原子核外电子运动的近代概念 （2）四个量子数,s、p、d原子轨道的形状和伸展方向 （3）原子核外电子排布的一般规律 （1）modern concept of electron movement outside the nucleus （2）four quantum numbers, s, p, d atomic orbital shapes, and their extension directions （3）the general rule of electron arrangement outside the nucleus	2
1.3	多电子原子轨道能级与周期性 orbit energy level and periodicity of multi-electron atoms	（1）电子层结构特征 （2）结构和性质的关系 （3）电离能、电子亲和能、电负性及主要氧化值的周期性变化规律 （1）properties of electronic layer structure （2）relationship between structure and properties （3）periodic changes of ionization energy, electron affinity energy, electronegativity, and main oxidation value	2

第二章 分子结构（molecular structure）

章节序号 chapter number	章节名称 chapters	知识点 key points	课时 class hour
2.1	离子键的形成和特点 formation of ionic bond and its characteristics	（1）化学键的特征和类型 （2）离子键的形成和特点、离子极化、晶格能 （1）characteristics and types of chemical bonds （2）formation of ionic bond and its characteristics, ionic polarization, lattice energy	3
2.2	共价键与分子构型 covalent bond and molecular configuration	（1）价键理论 （2）杂化轨道理论 （3）价层电子对互斥理论 （4）分子轨道理论 （1）valence bond theory （2）hybrid orbital theory （3）valence shell electron pair repulsion theory （4）molecular orbital theory	6
2.3	金属键 metal bond	（1）自由电子模型 （2）能带理论 （1）free electron model （2）band theory	1

第三章　宏观物质及其聚集状态（macroscopic matter and its aggregation state）

章节序号 chapter number	章节名称 chapters	知识点 key points	课时 class hour
3.1	分子间作用力 intermolecular force	（1）分子间力 （2）氢键 （1）intermolecular force （2）hydrogen bond	1
3.2	气体 gas	（1）理想气体 （2）实际气体 （3）范德华气体方程 （1）ideal gas （2）actual gas （3）van der waals gas equation	1
3.3	液体 liquid	（1）气体的液化 （2）液体的蒸发和凝固 （1）liquefaction of gas （2）evaporation and solidification of liquid	1
3.4	晶体与非晶体 crystal and amorphous	（1）晶体结构和空间点阵 （2）晶体的基本类型 （3）非晶体 （1）crystal structure and spatial lattice （2）the basic types of crystals （3）amorphous	1

第四章　溶液与胶体（solution and colloid）

章节序号 chapter number	章节名称 chapters	知识点 key points	课时 class hour
4.1	溶液 solution	（1）溶液的浓度 （2）理想溶液 （1）concentration of the solution （2）ideal solution	1
4.2	胶体 colloid	（1）胶体的结构、制备、性质 （2）溶胶的稳定性和聚沉作用 （1）the structure, preparation and properties of colloid （2）stability and coagulation of colloid	1

第五章 热力学第一定律（the first law of thermodynamics）

章节序号 chapter number	章节名称 chapters	知识点 key points	课时 class hour
5.1	热力学基本概念 basic concepts of thermodynamics	（1）基本概念 （2）热力学第一定律 （1）basic concepts （2）the first law of thermodynamics	1
5.2	体积功与可逆过程 expansion work and reversible change	（1）体积功 （2）可逆过程 （1）expansion work （2）reversible change	1
5.3	热效应与焓 heat transactions and enthalpy	（1）等容热和等压热 （2）焓的定义 （1）heat at constant volume and heat at constant pressure （2）the definition of enthalpy	1
5.4	绝热过程与卡诺循环 adiabatic changes and Carnot cycle	（1）绝热可逆方程 （2）卡诺循环 （1）equation of reversible adiabatic （2）Carnot cycle	1
5.5	热化学 thermochemistry	（1）热化学 （2）标准摩尔反应热效应的计算基础 （1）thermochemistry （2）calculation basis of heat effect for standard molar reaction	2

第六章 热力学第二定律（the second law of thermodynamics）

章节序号 chapter number	章节名称 chapters	知识点 key points	课时 class hour
6.1	熵 entropy	（1）热力学第二定律 （2）卡诺定律 （3）熵函数及熵增原理 （4）熵变的计算 （1）the second law of thermodynamics （2）Carnot law （3）entropy function and principle of increase of entropy （4）calculation of entropy change	2

续表

章节序号 chapter number	章节名称 chapters	知识点 key points	课时 class hour
6.2	熵的统计学意义及热力学第三定律 the statistical significance of entropy and the third law of thermodynamics	（1）热力学第三定律 （2）规定熵 （1）the third law of thermodynamics （2）prescribed entropy	1
6.3	亥姆霍兹能和吉布斯能 Helmholtz energy and Gibbs energy	（1）亥姆霍兹能和吉布斯能 （2）吉布斯能的计算 （3）化学反应标准吉布斯能 （1）Helmholtz energy and Gibbs energy （2）calculation of Gibbs energy （3）standard reaction Gibbs energies	2
6.4	热力学第一定律与第二定律的统一 combining the first and the second laws	（1）热力学基本方程式 （2）麦克斯韦关系式 （1）fundamental equation of thermodynamics （2）Maxwell equations	1
6.5	偏摩尔量和化学势 partial molar quantities and chemical potential	（1）偏摩尔量 （2）化学势 （3）化学势的意义 （1）partial molar quantities （2）chemical potential （3）the wider significance of the chemical potential	1
6.6	化学势的表达与应用 expression and application of chemical potential	（1）纯物质的化学势 （2）溶液中各组分的化学势 （3）溶液的依数性 （1）chemical potential of pure substances （2）chemical potential of each component in the solution （3）colligative properties	3

第七章 化学平衡（chemical equilibrium）

章节序号 chapter number	章节名称 chapters	知识点 key points	课时 class hour
7.1	化学反应等温方程式 isothermal equation for chemical reactions	（1）化学反应的平衡条件 （2）标准反应吉布斯能变化 （1）equilibrium conditions of chemical reactions （2）standard reaction Gibbs energy change	1

续表

章节序号 chapter number	章节名称 chapters	知识点 key points	课时 class hour
7.2	化学反应平衡常数 equilibrium constant of chemical reactions	（1）平衡常数的表达 （2）平衡常数的计算 （1）expression of equilibrium constants （2）calculation of equilibrium constants	1
7.3	影响化学平衡的因素 influencing factors of equilibrium constant	（1）温度对平衡常数的影响 （2）压力对平衡的影响 （3）其他因素对化学平衡的影响 （1）the response of equilibria to temperature （2）how equilibria respond to pressure （3）the other factors on chemical balance	2

第八章 化学动力学基础（fundamentals of chemical kinetics）

章节序号 chapter number	章节名称 chapters	知识点 key points	课时 class hour
8.1	化学反应速率 rates of chemical reactions	（1）化学反应速率的定义 （2）基元反应 （3）反应级数和速率常数 （1）definition of rate （2）elementary reaction （3）reaction order and rate constants	2
8.2	简单级数反应 simple order reactions	（1）一级反应 （2）二级反应 （3）反应级数的测定法 （1）the first order reactions （2）the second order reactions （3）determination of reaction order	2
8.3	阿伦尼乌斯公式 Arrhenius equation	（1）温度对反应速率的影响 （2）阿伦尼乌斯公式及活化能 （1）the temperature dependence of reaction rates （2）Arrhenius equation and activation energy	1
8.4	催化剂简介 introduction to catalyst	（1）催化作用 （2）均相催化和多相催化 （1）catalysis （2）homogeneous and heterogeneous catalysis	1

第九章 离子型反应（ionic reactions）

章节序号 chapter number	章节名称 chapters	知识点 key points	课时 class hour
9.1	沉淀反应 precipitation reaction	（1）难溶电解质溶度积 （2）溶度积规则及应用 （3）典型沉淀反应与储能材料 （1）solubility product constant of insoluble electrolyte （2）rule of solubility product and its application （3）typical precipitation reactions and energy storage materials	2
9.2	配位反应 coordination reaction	（1）配位平衡 （2）配位反应机理 （3）配合物与储能材料 （1）coordination equilibrium （2）mechanism of coordination reaction （3）coordination compounds for energy storage materials	4
9.3	酸碱反应 acid and base reaction	（1）酸碱质子理论 （2）酸碱平衡 （3）酸碱平衡移动 （4）缓冲溶液 （1）acid-base proton theory （2）acid-base equilibrium （3）shift of acid-base equilibrium （4）buffer solution	5
9.4	亲电反应 electrophilic reaction	（1）亲电加成反应 （2）亲电取代反应 （3）亲电加成及取代反应与能源存储 （1）electrophilic addition reaction （2）electrophilic substitution reaction （3）electrophilic addition and substitution reactions in energy storage	8
9.5	亲核反应 nucleophilic reaction	（1）亲核取代反应 （2）亲核加成反应 （3）亲核加成及取代反应与能源存储 （1）nucleophilic substitution reaction （2）nucleophilic addition reaction （3）nucleophilic addition and substitution reactions in energy storage	5

第十章 自由基反应和聚合反应（radical reaction and polymerization）

章节序号 chapter number	章节名称 chapters	知识点 key points	课时 class hour
10.1	自由基反应 radical reaction	（1）自由基的基本概念 （2）自由基的种类 （3）自由基的稳定性 （4）自由基的活性 （5）自由基的反应 （1）introduction of radical （2）the type of radical （3）the stability of radical （4）the reactivity of radical （5）the reaction of radical	2
10.2	聚合反应 polymerization	（1）高分子简介 （2）逐步聚合 （3）自由基聚合 （4）离子聚合 （5）配位聚合 （6）开环聚合 （1）introduction of polymers （2）step-growth polymerization （3）radical polymerization （4）ionic polymerization （5）coordination polymerization （6）ring-opening polymerizations	6

第十一章 氧化还原反应（redox reaction）

章节序号 chapter number	章节名称 chapters	知识点 key points	课时 class hour
11.1	氧化还原反应概述 overview of redox reaction	（1）氧化值 （2）氧化和还原反应 （3）氧化还原反应的配平 （4）氧化还原反应的标准平衡常数 （1）oxidation number （2）oxidation and reduction reaction （3）balance of redox reaction （4）standard equilibrium constant of redox reaction	2

续表

章节序号 chapter number	章节名称 chapters	知识点 key points	课时 class hour
11.2	原电池 primary battery	（1）原电池的组成 （2）电极及其分类 （3）原电池的符号 （1）composition of primary battery （2）electrode and its classification （3）symbol of primary battery	2
11.3	电极与电极电势 electrode and electrode potential	（1）电极电势的产生 （2）标准电极电势 （3）Nernst方程 （4）电极电势的应用 （1）generation of electrode potential （2）standard electrode potential （3）Nernst equation （4）application of electrode potential	4
11.4	常用化学电池 typical chemical batteries	（1）锌锰电池 （2）锂电池 （3）燃料电池 （1）Zn-Mn batteries （2）lithium battery （3）fuel cells	2

第十二章　能源化学基础（fundamentals of energy chemistry）

章节序号 chapter number	章节名称 chapters	知识点 key points	课时 class hour
12.1	能源化学概述 overview of energy chemistry	（1）能源的分类 （2）能量的转化 （1）classification of energy sources （2）conversion of energy	0.5
12.2	不可再生能源 non-renewable energy resource	（1）化石能源 （2）核能 （1）fossil energy （2）nuclear energy	0.5
12.3	可再生能源 renewable energy resource	（1）化学电源 （2）氢能 （3）太阳能 （4）生物质能	

续表

章节序号 chapter number	章节名称 chapters	知识点 key points	课时 class hour
12.3	可再生能源 renewable energy resource	（1）electrochemical power source （2）hydrogen energy （3）solar energy （4）biomass energy	1

第十三章 含能物质基础 （fundamentals of energetic materials）

章节序号 chapter number	章节名称 chapters	知识点 key points	课时 class hour
13.1	含能物质概述 overview of energetic materials	（1）常见含能物质的分类 （2）含能物质的应用 （1）classification of common energetic substances （2）application of energetic substances	0.5
13.2	含能物质的合成方法 energetic material synthesis method	（1）氢气的合成方法 （2）铝粉的合成方法 （3）常用叠氮类物质的合成方法 （4）硝化棉/硝化甘油的合成方法 （1）hydrogen synthesis method （2）synthesis method of aluminum powder （3）synthetic method of commonly used azide substances （4）synthesis method of nitrocellulose/nitroglycerin	1
13.3	含能物质的热力学 thermodynamics of energetic matter	（1）含能材料的热力学行为 （2）含能材料的热力学性质 （3）高能燃料的选取原则 （1）thermodynamic behavior of energetic materials （2）thermodynamic properties of energetic materials （3）selection principle of high-energy fuel	0.5

第十四章 储能化学相关材料简介（introduction of energy storage chemistry related materials）

章节序号 chapter number	章节名称 chapters	知识点 key points	课时 class hour
14.1	储氢材料 hydrogen storage materials	（1）储氢材料的分类 （2）储氢材料的基本特征 （3）储氢材料的发展概况 （1）classification of hydrogen storage materials （2）characteristics of hydrogen storage materials （3）development of hydrogen storage materials	0.5

续表

章节序号 chapter number	章节名称 chapters	知识点 key points	课时 class hour
14.2	电催化剂材料 electrocatalyst materials	（1）电催化材料的分类 （2）电催化材料的基本特征 （3）电催化材料的应用概况 （1）classification of electrocatalytic materials （2）characteristics of electrocatalytic materials （3）application of electrocatalytic materials	0.5
14.3	锂电池材料 lithium battery materials	（1）锂电池的工作原理 （2）锂电池的电极材料 （3）锂电池的隔膜材料 （1）principle of lithium batteries （2）electrode materials of lithium batteries （3）separator of lithium batteries	0.5
14.4	电容器材料 capacitor materials	（1）电容器的工作原理 （2）电容器的电极材料 （3）电容器的隔膜材料 （1）principle of capacitor （2）electrode materials of capacitor （3）separator of capacitor	0.5

4.3.5 实验环节（Experiments）

序号 number	实验内容 experiment content	知识点 key points	课时 class hour
1	实验室安全及酸碱滴定 laboratory safety and acid-base titration	（1）实验室安全基本知识 （2）差减法称量 （3）溶液的配制 （4）酸碱滴定 （1）basic knowledge of laboratory safety （2）addition and subtraction of measuring mass （3）preparation of solution （4）acid-base titration	4
2	硫酸亚铁铵的制备 preparation of ammonium ferrous sulfate	无机物制备及含量检测 preparation and content detection of inorganics	4

续表

序号 number	实验内容 experiment content	知识点 key points	课时 class hour
3	邻二氮菲分光光度法测定微量铁 spectrophotometric determination of trace iron with o-phenanthroline method	（1）分光光度法测定样品含量 （2）分光光度计的使用、标准曲线法 （1）determination of the contents of the sample by spectrophotometric （2）applications of spectrophotometer, standard curve method	4
4	金溶胶的制备及其光学性质与界面自组装 preparation, optical properties, and interface self-assembly of gold colloid	（1）纳米材料的制备 （2）光学性质检测 （3）界面自组装原理及应用 （1）preparation of nanomaterials （2）detection of optical properties （3）principle and application of interface self-assembly	4
5	化学反应速率常数和活化能的测定 determination of rate constant and activation energy of chemical reactions	（1）浓度对化学反应速率的影响 （2）温度对化学反应速率的影响 （3）催化剂对化学反应速率的影响 （1）effect of concentration on the reaction rate （2）effect of temperature on the reaction rate （3）effect of catalyst on chemical reaction rate	4
6	熔点、沸点及折光率的测定 determination of the melting/boiling point and refractive index	（1）蒂勒管使用 （2）阿贝折光仪的使用 （3）熔沸点的测定方法 （1）use of Thiele melting point tube （2）use of Abbe refractometer （3）method for determination of melting and boiling point	4
7	环己烯的制备 preparation of cyclohexene	（1）样品分馏、分液 （2）液体样品干燥 （1）fractionation and separation of samples （2）drying of liquid samples	4
8	溴乙烷的制备 preparation of bromoethane	低沸点化合物常压蒸馏原理及方法 principle and method of atmospheric distillation of compounds with low boiling point	4
9	乙酰水杨酸（阿司匹林）的制备 preparation of acetylsalicylic acid	（1）药物合成 （2）样品重结晶、抽滤 （1）synthesis of drug （2）recrystallization and extraction of samples	4

续表

序号 number	实验内容 experiment content	知识点 key points	课时 class hour
10	苯乙烯与丙烯腈溶液共聚及竞聚率的测定 solution copolymerization of styrene and acrylonitrile and determination of reactivity ratio	（1）单体的精制与共聚 （2）产物的提纯、结构表征和竞聚率计算 （1）refining and copolymerization of monomers （2）purification, structure characterization and reactivity ratio calculation of the product	4
11	液体饱和蒸汽压的测定 determination of saturated vapor pressure of liquids	（1）实验通过减压沸腾的方法测定不同温度下纯乙醇的饱和蒸汽压 （2）通过数据拟合计算乙醇的摩尔蒸发焓和正常沸点 （1）the saturated vapor pressure of pure ethanol at different temperatures was measured by vacuum boiling met （2）the molar enthalpy of evaporation and normal boiling point of ethanol calculated by data fitting	4
12	锂离子电池材料回收虚拟仿真实验 virtual simulation experiment of recycling of lithium-ion battery material	锂离子电池材料回收流程的设计、回收过程参数仿真优化 design of the recycling process of lithium-ion battery materials and simulation optimization of recovery process parameters	4

4.4 "计算机科学基础与高级程序设计"教学大纲

课程名称：计算机科学基础与高级程序设计

Course Title：Computer Science Fundamentals and Advanced Programming Design

先修课程：无

Prerequisites：None

学分：6

Credits：6

4.4.1　课程目的和基本内容（Course Objectives and Basic Contents）

本课程是面向储能科学与工程专业的计算机基础课程，属于数学和基础科学类必修课。

课程第一部分主要讲授离散数学，包括集合论和图论两部分内容。这部分的学习将为储能科学与工程专业学生奠定后续计算机类课程学习的数学基础。

课程第二部分主要讲授内容包括 C++ 程序设计和数据结构与算法两大部分。

C++ 程序设计部分，主要结合 C/C++ 语言，对程序设计的基本理论、面向过程和面向对象程序设计方法展开讨论，同时介绍一些基本算法。具体内容如下： C++ 程序设计的数据结构，包括整型、浮点、字符、数组、指针、结构体等； C++ 的控制结构，包括顺序、分支和循环结构；函数的声明、定义、调用、重载以及模板；C++ 面向对象编程的基本概念、思想和方法，包括类和对象的定义、类的构造函数和析构函数以及类运算符的重载。

数据结构与算法部分，主要通过对数据结构基础理论、基本数据结构及应用算法展开讨论。具体内容包括：数据结构基础理论，包括概念、逻辑结构、存储结构，抽象数据类型、算法和算法分析等；基本数据结构的定义、表示方法、实现和应用，包括线性表、栈、队列、树、二叉树、图等；数据结构应用算法的思想、过程、实现方法、复杂度分析，包括排序、搜索等。

This course is a basic computer course for students majoring in energy storage science and engineering and is a mandatory course in the category of mathematics and basic science.

The first part of this course focuses on discrete mathematics, including set theory and graph theory. The study of this part will lay the mathematical foundation for the follow-up computer courses for students majoring in energy storage science and engineering.

The second part of the course focuses on C++ programming and data structures along with related algorithms.

The C++ programming part discusses the basic theories and the design methods of programming, procedure-oriented programming, and object-oriented programming, along with some basic algorithms. In detail, this part includes the data structures of C++ programming, including integer, float, character, array, pointer, and structure. Furthermore, it introduces the basic control structures of C++, including sequence, branch, and loop. The declaration, definition, invoke, overload, and template of functions are discussed. The basic concepts, ideas, and methods of object-oriented programming including the concept of class, object, constructor, destructor as well as operators overloading are introduced.

The data structures and algorithms discuss the fundamental theories of data structures, basic data structures, and application algorithms. In detail, this part includes the fundamental theories, including data structure concepts, data logical structure and storage structure, abstract data type, algorithm, performance analysis, and so on; the definition, representation method, implementation, and applications of basic data structures, including linear lists, stacks, queues, trees, binary trees, graphs; the basic idea, process, implementation, complexity analysis of application algorithms, including sorting and search.

4.4.2　课程基本情况（Basic Information of the Course）

课程名称	计算机科学基础与高级程序设计 Computer Science Fundamentals and Advanced Programming Design								
开课时间	一年级		二年级		三年级		四年级		总学分
	秋	春	秋	春	秋	春	秋	春	

开课时间	一年级	二年级	三年级	四年级	总学分	6	总学时	104
	秋　春	秋　春	秋　春	秋　春				
课程定位	储能科学与工程专业本科生储能基础课程群必修课							
授课学时分配	课堂讲授96学时+实验8学时							
先修课程	无							
后续课程	嵌入式智能系统,物联网应用概论,工程图学							
教学方式	课堂教学、平时作业、阶段测试							
考核方式	第一部分:课程结束笔试成绩占65%,阶段测试占20%,平时成绩占15% 第二部分:课程期末考试成绩占60%,实验占30%,平时成绩占10%							
参考教材	[1] 陈建明,曾明,刘国荣.离散数学. 3版. 西安:西安交通大学出版社,2012. [2] 普拉达. C++ Primer Plus.影印版. 北京:人民邮电出版社,2015. [3] 普拉达. C++ Primer Plus. 6版. 中文版. 张海龙,袁国忠,译. 北京:人民邮电出版社,2012. [4] Shaffer C A. 数据结构与算法分析:C++版.2版. 北京:电子工业出版社,2009.							
参考资料	[1] Kolman B, Busby R C,Ross S C. Discrete Mathematical Structures.北京:高等教育出版社,2010. [2] 刘国荣,李文,陈建明. 离散数学内容提要与习题解析.西安:西安交通大学出版社,2014.							

4.4.3　教学目的和基本要求（Teaching Objectives and Basic Requirements）

（1）有效地掌握本课程中的所有概念，包括集合、关系、函数、图等，并能综合应用于解决储能领域的复杂工程问题。

（2）通过讲授课程中大量定理的证明过程、复习概念、精讲各种证明方法，使学生在了解定理的内容和结论的基础上，掌握数学证明中常用的方法，例如归纳法、反证法、二分法、枚举法（穷举法）等，特别着重于存在性、结构性、构造性证明方法，并能够综合运用证明方法对储能领域新的问题进行建模、推理和论证。

（3）通过各类证明方法的学习，启发学生独立思维的能力，并进一步创造性地提出解决问题的方法，锻炼和提高学生解决问题的能力，培养学生严谨的逻辑思维方法。

（4）应能掌握基本的编程思想，能够使用 C++ 语言编写程序，能够通过编写程序完成储能系统复杂科学与工程问题的模拟、分析、求解和论证。

（5）培养学生数据抽象能力、适当的逻辑结构及存储结构选择能力，以及设计算法解决一般应用问题的能力。

（1）Effectively master all the concepts in this course, including set, relation, function, graph, etc., and can comprehensively apply these concepts to solve complex engineering problems in the field of energy storage.

（2）By studying the proof process of a large number of theorems in the course, reviewing concepts, and elaborating various proof methods, students can master the commonly used methods in mathematical proof based on understanding the content and conclusions of the theorems, such as induction, counter-proof, dichotomy method, enumeration method (exhaustion method), etc., with particular emphasis on the existence, structural and constructive proof methods, and can comprehensively use the proof method to model, reason and demonstrate new problems in the field of energy storage.

（3）Through the study of various proof methods, students' independent thinking ability is inspired, and students can further creatively propose their own problem-solving methods. Exercise and improve students' problem-solving skills, and cultivate students' rigorous logical thinking methods.

（4）Students should be able to master basic programming ideas, be able to use C++ language to write programs, and be able to complete the simulation, analysis, solving, and demonstration of complex scientific and engineering problems of energy storage systems through programming.

（5）Cultivate students' ability of data abstraction, selecting appropriate logical structure and storage structure, and designing algorithms to solve general application problems.

4.4.4 课程大纲和知识点（Syllabus and Key Points）

第一部分

第一章 集合（set）

章节序号 chapter number	章节名称 chapters	知识点 key points	课时 class hour
1.1	集合的基本概念与性质 basic concept and properties of set	（1）集合、子集、空集、幂集 （2）集合的性质 （1）set, subset, empty set, power set （2）properties of set	2
1.2	集合的基本运算 basic operations of set	（1）交、并、补运算 （2）相关定理 （1）intersection, union, and complement operations （2）related theorems	2

续表

章节序号 chapter number	章节名称 chapters	知识点 key points	课时 class hour
1.3	集合的宏运算 macro operations on set	（1）差、对称差运算 （2）相关定理 （1）difference, symmetric difference operation （2）related theorems	1

第二章　关系（relation）

章节序号 chapter number	章节名称 chapters	知识点 key points	课时 class hour
2.1	集合的叉积和元组 cross product and tuple of set	叉积和元组 cross product and tuple	1
2.2	关系的基本概念、表示和性质 basic concept, representations and properties of relation	（1）基本概念和表示 （2）关系的性质 （1）concept, representations （2）properties of relation	2
2.3	关系的运算、复合幂及闭包 relation operations, compound powers, and closures	（1）逆运算和复合运算 （2）复合幂和闭包 （1）inverse operation and compound operation （2）compound powers and closures	2
2.4	等价关系 equivalence relation	（1）等价关系的基本概念、性质 （2）等价类和划分 （1）basic concept and properties of equivalence relation （2）equivalence classes and partition	2
2.5	半序关系 partial order relation	（1）半序关系的基本概念、性质 （2）半序、全序、良序 （3）哈塞图 （1）basic concept and properties of partial order relation （2）partial order, linear order and well order （3）Hasse	2

第三章　函数（functions）

章节序号 chapter number	章节名称 chapters	知识点 key points	课时 class hour
3.1	函数的基本概念与性质 basic concept and properties of functions	（1）函数的基本概念和性质 （2）逆函数、复合函数、置换 （1）basic concept and properties of functions （2）inverse function, composite function, and permutation	2
3.2	集合的基数 cardinality of set	（1）等势 （2）可数集、不可数集 （1）set equipotentiality （2）countable set, uncountable set	2

第四章　图论（graph theory）

章节序号 chapter number	章节名称 chapters	知识点 key points	课时 class hour
4.1	图的基本概念 basic concept of graph	（1）图的概念、图的可达和连通 （2）路与圈 （3）图的矩阵表示 （1）concept, reachability, and connectivity of graph （2）path and circle （3）matrix representation of graph	4
4.2	带权图的最短路径 shortest path with weighted graph	迪杰斯特拉算法 Dijkstra algorithm	2
4.3	欧拉图和哈密顿图 Euler graph and Hamilton graph	（1）欧拉图和欧拉定理 （2）哈密顿图和相关定理 （1）Euler graph and Euler theorem （2）Hamilton graph and related theorems	3
4.4	二分图和平面图 bipartite graph and planar graph	（1）二分图和最大匹配 （2）平面图、欧拉公式推论、K技术 （1）bipartite graph and maximum matching （2）planar graph, Euler formula inference, K technique	3
4.5	树 tree	（1）树的概念和性质 （2）生成树和最优生成树 （1）concept and properties of tree （2）spanning tree and optimal spanning tree	2

第二部分

第一章　绪论（introduction）

章节序号 chapter number	章节名称 chapters	知识点 key points	课时 class hour
1.1	绪论 introduction	（1）计算机编程语言的发展 （2）面向过程和面向对象编程的比较 （3）C++程序开发工具 （1）history of computer programming language （2）comparison of procedure-oriented and object-oriented programming （3）tools for C++ programming	2

第二章　C++ 基本知识（fundamentals of C++ programming）

章节序号 chapter number	章节名称 chapters	知识点 key points	课时 class hour
2.1	C++基本知识 C++ initials	（1）main函数 （2）注释和代码格式化 （3）C++头文件、预处理器和iostream文件 （4）cout进行输出 （1）main function （2）comments and source code formatting （3）C++ header file, preprocessor, and the iostream file （4）output with cout	2
2.2	C++语句 C++ statements	（1）声明语句和变量 （2）赋值语句 （3）cin和cout （1）declaration statements and variables （2）assignment statements （3）cin and cout	1
2.3	函数 functions	（1）带返回值的函数 （2）函数变体 （1）a function with a return value （2）function variations	1

第三章 数据处理（data and operations）

章节序号 chapter number	章节名称 chapters	知识点 key points	课时 class hour
3.1	简单变量 simple variables	（1）变量名 （2）整型short、int、long和long long （3）无符号类型 （4）常量类型 （5）char类型 （6）bool类型 （1）names for variables （2）short, int, long, and long long integer types （3）unsigned types （4）constants types （5）the char type （6）the bool type	0.5
3.2	浮点数 floating-point numbers	（1）浮点数 （2）浮点类型 （3）浮点常量 （1）floating-point number （2）floating-point types （3）floating-point constants	0.5
3.3	C++算术运算符 C++ arithmetic operators	（1）运算符的优先级和结合性 （2）除法运算符 （3）取模 （4）类型转换 （1）order of operation: operator precedence and associativity （2）division operator （3）the modulus operator （4）type conversions	1

第四章 复合数据类型（compound data）

章节序号 chapter number	章节名称 chapters	知识点 key points	课时 class hour
4.1	数组 arrays	（1）数组介绍 （2）数组初始化规则 （1）introduction of arrays （2）initialization rules for arrays	1

续表

章节序号 chapter number	章节名称 chapters	知识点 key points	课时 class hour
4.2	字符串 strings	（1）拼接字符串常量 （2）数组中使用字符串 （3）字符串输入 （1）concatenating string literals （2）using strings in an array （3）string input	1
4.3	string类 string class	（1）string 类 （2）赋值、拼接和附加以及其他操作 （3）string 类 I/O （1）string class （2）assignment, concatenation, and appending （3）I/O of string class	1
4.4	结构简介 introduction of structures	（1）在程序中使用结构 （2）结构属性 （3）结构数组 （1）using a structure in a program （2）properties of structures （3）arrays of structures	1
4.5	指针和自由存储空间 pointers and the free store	（1）声明和初始化指针 （2）指针和数字 （3）使用 new 和 delete （1）declaring and initializing pointers （2）pointers and numbers （3）allocating memory using new and freeing memory with delete	2

第五章　循环控制结构（loop structure）

章节序号 chapter number	章节名称 chapters	知识点 key points	课时 class hour
5.1	for语句 for loops	（1）for 循环的组成部分 （2）++ 和 -- （3）复合语句 （4）关系表达式 （5）冒泡排序	1

续表

章节序号 chapter number	章节名称 chapters	知识点 key points	课时 class hour
5.1	for 语句 for loops	（1）parts of a for loop （2）++ and -- （3）compound statements, or blocks （4）relational expressions （5）bubble sort algorithm	1
5.2	while循环和do while循环 while loop and do while loop	（1）for 与 while （2）do while 循环 （3）循环和文本输入 （4）嵌套循环和二维数组 （1）for and while （2）do while loop （3）loops and text input （4）nested loops and two-dimensional arrays	2

第六章 分支控制结构（branching structure）

章节序号 chapter number	章节名称 chapters	知识点 key points	课时 class hour
6.1	if 语句 if statement	（1）if else 语句 （2）格式化 if else 语句 （3）if else if else语句 （1）if else statement （2）formatting if else statement （3）if else if else statement	1
6.2	逻辑表达式 logical expressions	（1）逻辑运算符 （2）字符函数库 cctype （3）?：运算符 （4）switch语句 （5）break 和 continue 语句 （1）logical operator （2）cctype library of character functions （3）?：operator （4）switch statement （5）break and continue statements	1

第七章　函数（functions）

章节序号 chapter number	章节名称 chapters	知识点 key points	课时 class hour
7.1	函数的基本知识 function review	（1）定义函数 （2）函数原型和函数调用 （1）defining a function （2）prototyping and calling a function	0.5
7.2	函数参数和按值传递 function arguments and passing by value	（1）多个参数 （2）接受两个参数值传递 （1）multiple arguments （2）two-argument function	1
7.3	函数和数组 functions and arrays	（1）函数如何使用指针来处理数组 （2）使用数组区间的函数 （3）函数和二维数组 （1）how pointers enable array-processing functions （2）functions using array ranges （3）functions and two-dimensional arrays	1
7.4	函数和C-风格字符串 functions and C-style strings	（1）C-风格字符串作为参数 （2）返回C-风格字符串 （1）functions with C-style string arguments （2）functions that return C-style strings	0.5
7.5	递归 recursion	（1）单一递归调用 （2）多重递归调用 （1）recursion with a single recursive call （2）recursion with multiple recursive calls	1

第八章　函数进阶（advances in functions）

章节序号 chapter number	章节名称 chapters	知识点 key points	课时 class hour
8.1	C++内联函数 C++ inline functions	内联函数使用 using inline functions	0.5
8.2	引用变量 reference variables	（1）创建引用变量 （2）引用和类 （3）对象、继承和引用 （4）何时使用引用 （1）creating a reference variable （2）using reference with a class object （3）objecs, inheritance, and reference （4）when to use reference arguments	1

续表

章节序号 chapter number	章节名称 chapters	知识点 key points	课时 class hour
8.3	函数重载 function overloading	（1）函数重载 （2）何时使用函数重载 （1）function overloading （2）when to use function overloading	0.5

第九章 对象和类（objects and classes）

章节序号 chapter number	章节名称 chapters	知识点 key points	课时 class hour
9.1	过程性编程和OO编程 procedural and OO programming	过程性编程和OO编程 procedural and OO programming	0.5
9.2	抽象和类 abstraction and classes	（1）C++中的类 （2）实现类成员函数 （3）使用类 （1）classes in C++ （2）implementing class member functions （3）using classes	1
9.3	构造和析构函数 class constructors and destructors	（1）构造函数 （2）析构函数 （3）构造函数和析构函数小结 （1）constructors （2）destructors （3）conclusion of constructors and destructors	1.5

第十章 使用类（working with classes）

章节序号 chapter number	章节名称 chapters	知识点 key points	课时 class hour
10.1	运算符重载 operator overloading	运算符重载 operator overloading	2
10.2	运算符重载实例 developing an operator overloading examples	（1）加法运算符 （2）重载限制 （3）重载其他运算符 （1）adding an additional operator （2）overloading restrictions （3）more overloading operators	2

第十一章　数据结构预备知识（preliminaries of data structures）

章节序号 chapter number	章节名称 chapters	知识点 key points	课时 class hour
11.1	数据结构 data structure	（1）基本概念 （2）逻辑结构和存储结构之间的关系 （1）basic concepts of terms （2）relationship between logical structure and storage structure	0.5
11.2	抽象数据类型的表示与实现 representation and implementation of abstract data types	（1）抽象数据类型的定义 （2）抽象数据类型的表示 （3）抽象数据类型的实现方法 （1）definitions of abstract data type （2）representation of abstract data type （3）implementation of abstract data type	0.5
11.3	算法和算法分析 algorithm and performance analysis	（1）算法五要素 （2）时间复杂度的计算方法 （1）five components of algorithms （2）computational method of time complexity	1

第十二章　线性表（linear lists）

章节序号 chapter number	章节名称 chapters	知识点 key points	课时 class hour
12.1	线性表 linear list	（1）线性表的定义 （2）线性表的顺序表示和实现 （3）线性表的链式表示和实现 （1）definition of linear list （2）array representation and implementation of a linear list （3）linked representation and implementation of a linear list	2
12.2	栈 stacks	（1）栈的定义 （2）顺序栈、链式栈实现 （3）栈与递归 （1）definition of stacks （2）implementation of array stacks and linked stacks （3）stacks and recursion	2
12.3	队列 queues	（1）队列的定义 （2）循环队列、链式队列实现 （1）definition of queues （2）implementation of circular queues and linked queues	2

第十三章　树和二叉树（trees and binary trees）

章节序号 chapter number	章节名称 chapters	知识点 key points	课时 class hour
13.1	树 trees	（1）一般树的定义 （2）一般树的遍历 （3）一般树的实现 （1）definition of general trees （2）traversal of general trees （3）implementation of general trees	2
13.2	二叉树 binary trees	（1）二叉树的定义与性质 （2）二叉树的遍历 （3）二叉树的结点实现 （1）definition and properties of binary trees （2）traversal of binary trees （3）implementation of binary trees	3
13.3	二叉树应用 applications of binary trees	（1）二叉搜索树 （2）堆与优先队列 （3）哈夫曼编码 （1）binary search trees （2）heap and priority queues （3）Huffman codes	3

第十四章　图（graph）

章节序号 chapter number	章节名称 chapters	知识点 key points	课时 class hour
14.1	图 graph	（1）图的定义和术语 （2）图的存储结构 （1）definition and terminology of graph （2）storage structure of graph	2
14.2	图的遍历 traversal of graph	（1）深度搜索 （2）广度搜索 （3）拓扑排序 （1）depth-first search （2）breadth-first search （3）topological sorting	2

第十五章 内排序（internal sorting）

章节序号 chapter number	章节名称 chapters	知识点 key points	课时 class hour
15.1	简单排序 simple sorting	（1）排序的基本概念 （2）插入排序 （3）冒泡排序 （4）选择排序 （1）concept of sorting （2）insertion sorting （3）bubble sorting （4）select sorting	4
15.2	高级排序 advanced sorting	（1）谢尔排序 （2）快速排序 （3）归并排序 （4）基数排序 （5）排序算法对比分析 （1）Shell sorting （2）quick sorting （3）merge sorting （4）radix sorting （5）comparison of sorting methods	4

第十六章 检索（search）

章节序号 chapter number	章节名称 chapters	知识点 key points	课时 class hour
16.1	静态查找 static search	（1）静态查找的概念 （2）顺序查找 （3）折半查找 （1）concept of static search （2）sequential search （3）binary search	2
16.2	散列表 Hash table	（1）哈希函数的构造方法 （2）处理冲突的方法 （3）哈希表的查找及其分析 （1）construction methods of Hash function （2）collision resolution （3）search of Hash table and its analysis	2

4.4.5　实验环节（experiments）

序号 number	实验内容 experiment content	知识点 key points	课时 class hour
1	学籍管理系统 student management system	（1）简单交互界面 （2）结构体数组和链表存储 （3）排序算法 （4）OO编程 （5）输入/输出 （1）simple interactive interface （2）structure array and link storage （3）sorting algorithm （4）object-oriented programming （5）input/output	4
2	二叉查找树编程与应用 implementation and application of binary search trees	（1）二叉查找树的建立 （2）中序遍历 （3）元素查找 （1）construct a binary search tree （2）inorder travel （3）element search	2
3	图的编程实现与应用 implementation and application of graphs	（1）图的建立 （2）深度搜索 （3）广度搜索 （4）拓扑排序 （1）construct a graph （2）depth-first search （3）breadth-first search （4）topological sorting	2

4.5　"自动控制理论"教学大纲

课程名称：自动控制理论

Course Title：Principles of Automatic Control

先修课程：线性代数与解析几何，复变函数与积分变换，大学物理（含实验），电路，现代电子技术

Prerequisites：Linear Algebra and Analytic Geometry，Functions of Complex Variable and Integral Transforms，Physics and Physics Experiments，Electric Circuit，Modern Electronic Technology

学分：3

Credits：3

4.5.1　课程目的和基本内容（Course Objectives and Basic Contents）

　　该课程是面向储能科学与工程专业本科三年级学生开设的专业大类基础课程，目的是让学生掌握自动控制的基本思想、基本概念和基本方法，从而为进一步深入学习其他专业课程奠定基础。

　　本课程的主要任务是使学生掌握自动控制系统的结构和工作原理，线性连续定常系统的微分方程、传递函数（方块图、信号流图）、频率特性以及状态空间表达式等数学模型；掌握线性连续定常系统的时域分析法、根轨迹分析法、频域分析法及串联校正装置的基本设计方法；掌握线性连续定常系统的李雅普诺夫稳定性分析、状态反馈和状态观测器的设计方法等。

　　This course is a fundamental course for third-year undergraduate students majoring in energy storage science and engineering. Its purpose is to enable students to master the basic ideas, basic concepts, and basic methods of automatic control, to lay the foundation for further in-depth study of other professional courses.

　　The main task of this course is to enable students to master the structure and operating principle of automatic control systems, mathematical models of differential equations, transfer functions (block diagram, signal flow diagram), frequency characteristics, and state-space expressions of linear continuous time-invariant systems. Master the time domain analysis method, root locus analysis method, frequency domain analysis method and the basic design method of series compensation devices of linear continuous time-invariant systems. Master the Lyapunov stability analysis, state feedback and state observer design method of linear continuous time-invariant systems, etc.

4.5.2　课程基本情况（Basic Information of the Course）

课程名称	自动控制理论 Principles of Automatic Control										
开课时间	一年级		二年级		三年级		四年级		总学分	3	总学时	56
	秋	春	秋	春	秋	春	秋	春				
课程定位	储能科学与工程专业本科生储能基础课程群必修课											
授课学时分配	课堂讲授48学时+实验8学时											
先修课程	线性代数与解析几何,复变函数与积分变换,大学物理（含实验）,电路,现代电子技术											
后续课程	储能系统检测与估计,信息物理融合能源系统											
教学方式	课堂教学、实验报告、平时作业											
考核方式	课程结束笔试成绩占70%,实验报告占10%,平时成绩占20%											

参考教材	[1] 张爱民. 自动控制原理. 2版. 北京:清华大学出版社,2019. [2] 杨清宇,马训鸣,朱洪艳,等. 现代控制理论. 2版. 西安:西安交通大学出版社,2020.
参考资料	无

4.5.3 教学目的和基本要求（Teaching Objectives and Basic Requirements）

（1）掌握自动控制的基本思想和概念以及自动控制系统的基本组成和分类，建立系统化思维方式；能够正确描述典型控制系统的工作原理，运用恰当的方法建立线性连续定常系统的数学模型，掌握传递函数和状态空间表达式之间的关系，掌握状态空间表达式的线性变换方法。

（2）结合典型控制系统，掌握稳定性、瞬态性能和稳态性能以及相应性能指标的理论和物理含义，熟练运用时域法、根轨迹法和频域法分析控制系统的稳定性、瞬态性能和稳态性能，掌握线性连续定常系统的状态空间分析方法，正确判断系统的状态能控性和能观测性，能够利用李雅普诺夫第二法分析线性控制系统的稳定性。

（3）能够运用控制系统的时域和频域性能指标，对控制系统性能进行评价。

（4）掌握控制系统的校正原理以及典型的超前、滞后和 PID 校正方法，熟悉典型校正环节与校正装置之间的对应关系，能够根据控制系统性能的综合评价结果，对问题进行识别，提出合理的控制器设计方法。掌握状态反馈极点配置和状态观测器的设计方法，掌握带有观测器的状态反馈控制系统设计方法。

（5）根据设计目标和任务，正确设计串联校正装置。

（6）正确运用 MATLAB 控制系统工具箱对控制系统进行建模、分析和设计。

（7）能够针对知识短板自主学习、查缺补漏，自觉关注控制理论与方法的新进展，勇于运用所学的控制知识解决实际工程问题。

（1）Master the basic ideas and concepts of automatic control and the basic composition and classification of automatic control systems, establish a systematic thinking mode. Be able to correctly describe the working principle of a typical control system, use appropriate methods to establish the mathematical model of a linear continuous time-invariant system, master the relationship between transfer functions and state-space expressions, and master the linear transformation method of state-space expressions.

（2）Combining typical control systems, master the theoretical and physical meanings of stability, transient and steady-state performances and corresponding performance indices, and proficiently use the time domain analysis method, root locus analysis method, and frequency domain analysis method to evaluate the stability, transient and steady-state performances of the control system. Master the state-space analysis method of linear continuous time-invariant systems, correctly judge the state controllability and observability of the system, and use Lyapunov's second method to analyze the stability of the linear control system.

(3) Be able to evaluate the performance of the control systems based on the time domain specifications and frequency domain response indices.

(4) Master the compensation principle of the control system and the typical lead, lag, and PID compensation methods. Be familiar with the relationship between the typical compensation method and the compensator. Be able to identify the problem according to the comprehensive evaluation results of the control system performance, and propose a reasonable controller design method. Master the state feedback and pole placement method and the state observer design method. Master the design of state feedback system with observer.

(5) According to the design objectives and tasks, correctly design the series compensator.

(6) Correctly use the control system toolbox of MATLAB to model, analyze, and design the control systems.

(7) Be able to learn independently, make up for shortcomings of knowledge, be aware of new developments and methods in control theory, and be willing to use the knowledge of control theory to solve practical engineering problems.

4.5.4　课程大纲和知识点（Syllabus and Key Points）

第一章　绪论（introduction）

章节序号 chapter number	章节名称 chapters	知识点 key points	课时 class hour
1.1	自动控制的基本概念 basic concepts of automatic control	自动控制的基本概念 basic concepts of automatic control	0.75
1.2	自动控制系统的基本形式 basic forms of automatic control systems	（1）开环控制系统 （2）闭环控制系统 （3）闭环控制系统的组成 （4）闭环控制系统的特点 （1）open-loop control systems （2）closed-loop control systems （3）composition of closed-loop control systems （4）characteristics of closed-loop control systems	0.25

续表

章节序号 chapter number	章节名称 chapters	知识点 key points	课时 class hour
1.3	自动控制系统分类 classification of automatic control systems	（1）按输入信号特征分类 （2）按系统中传递的信号的变化特征分类 （3）按系统特征分类 （4）按系统参数是否随时间变化分类 （1）classification by input signal characteristics （2）classification according to the change characteristics of transmitted signals in the system （3）classification by system characteristics （4）classification according to whether the system parameters change with time	0.5
1.4	对自动控制系统的基本要求 basic requirements for automatic control systems	对自动控制系统的基本要求 basic requirements for automatic control systems	0.5

第二章 控制系统的数学模型（mathematical model of control systems）

章节序号 chapter number	章节名称 chapters	知识点 key points	课时 class hour
2.1	微分方程 differential equations	（1）机械系统 （2）电路系统 （3）机电系统 （1）mechanical systems （2）electrical systems （3）electromechanical systems	1.25
2.2	传递函数 transfer functions	（1）传递函数定义 （2）典型环节传递函数 （3）举例说明建立传递函数的方法 （1）definition of transfer functions （2）typical link transfer functions （3）illustration of the method of establishing transfer functions	0.75
2.3	方块图 block diagrams	（1）方块图的组成和绘制 （2）方块图简化 （3）闭环系统的传递函数 （1）composition and drawing of block diagrams （2）simplification of block diagrams （3）transfer function of closed-loop systems	1

续表

章节序号 chapter number	章节名称 chapters	知识点 key points	课时 class hour
2.4	信号流图 signal-flow graph	（1）信号流图的组成和建立 （2）梅森增益公式 （1）composition and establishment of signal-flow graph （2）Mason's gain formula	1

第三章　控制系统的时域分析（time-domain analysis of control systems）

章节序号 chapter number	章节名称 chapters	知识点 key points	课时 class hour
3.1	引言 introduction	（1）典型输入信号及拉普拉斯变换 （2）瞬态响应和稳态响应 （3）瞬态性能指标和稳态性能指标 （1）typical input signal and Laplace transform （2）transient response and steady-state response （3）transient performance indices and steady-state performance indices	1
3.2	典型一阶系统的瞬态性能 transient performance of typical first-order systems	（1）一阶系统的数学模型 （2）一阶系统的单位脉冲响应 （3）一阶系统的单位阶跃响应 （4）一阶系统的单位斜坡响应 （5）一阶系统的单位加速度响应 （6）一阶系统的瞬态性能指标 （7）减小一阶系统时间常数的措施 （1）mathematical model of first-order systems （2）unit impulse response of first-order systems （3）unit step response of first-order systems （4）unit ramp response of first-order systems （5）unit acceleration response of first-order systems （6）transient performance indices of first-order systems （7）measures to reduce the time constant of first-order systems	1
3.3	典型二阶系统的瞬态性能 transient performance of typical second-order systems	（1）典型二阶系统的数学模型 （2）典型二阶系统的单位阶跃响应 （3）典型二阶系统的瞬态性能指标 （4）二阶系统瞬态性能的改善 （1）mathematical model of typical second-order systems （2）unit step response of typical second-order systems	2

续表

章节序号 chapter number	章节名称 chapters	知识点 key points	课时 class hour
3.3	典型二阶系统的瞬态性能 transient performance of typical second-order systems	（3）transient performance indices of typical second-order systems （4）improvement of transient performance of second-order systems	2
3.4	高阶系统的时域分析 time-domain analysis of higher-order systems	（1）三阶系统的瞬态响应 （2）高阶系统的瞬态响应 （3）闭环主导极点 （1）transient-response of third-order systems （2）transient-response of higher-order systems （3）closed-loop dominant pole	0.5
3.5	线性控制系统的稳定性分析 stability analysis of linear control systems	（1）线性控制系统的稳定性 （2）线性控制系统的稳定性的充要条件 （3）代数稳定性判据 （1）stability of linear control systems （2）necessary and sufficient conditions for stability of linear control systems （3）algebraic stability criteria	1.5
3.6	线性控制系统的稳态性能分析 steady-state performance analysis of linear control systems	（1）控制系统的误差和稳态误差 （2）稳态误差分析 （1）errors and steady-state errors of control systems （2）steady-state error analysis	2

第四章　控制系统的根轨迹 （root locus of control systems）

章节序号 chapter number	章节名称 chapters	知识点 key points	课时 class hour
4.1	引言 introduction	（1）根轨迹 （2）根轨迹的幅值和相角条件 （3）利用试探法确定根轨迹上的点 （1）root locus （2）amplitude and phase angle conditions of root locus （3）using trial method to determine the points on the root locus	0.5
4.2	绘制根轨迹的基本规则 basic rules of plotting root locus	（1）180° 等相角根轨迹的绘制规则 （2）0° 等相角根轨迹的绘制规则 （3）参量根轨迹	2

续表

章节序号 chapter number	章节名称 chapters	知识点 key points	课时 class hour
4.2	绘制根轨迹的基本规则 basic rules of plotting root locus	（4）关于180°和0°等相角根轨迹的几个问题 （1）plotting rules of 180°-equiphase root locus （2）plotting rules of 0°-equiphase root locus （3）parametric root locus （4）problems about 180° and 0°-equiphase root locus	2
4.3	控制系统根轨迹绘制实例 examples of root locus drawing of control systems	控制系统根轨迹绘制实例 examples of root locus drawing of control systems	0.5
4.4	基于根轨迹法的系统性能分析 system performance analysis based on root locus	（1）增加开环零、极点对根轨迹的影响 （2）条件稳定系统分析 （3）利用根轨迹估算系统的性能 （4）利用根轨迹计算系统的参数 （1）effect on root locus when adding zero and pole to open-loop （2）conditional stable system analysis （3）estimation of system performance using root locus （4）calculation of system parameters using root locus	1

第五章　控制系统的频域分析（frequency domain analysis of control systems）

章节序号 chapter number	章节名称 chapters	知识点 key points	课时 class hour
5.1	频率特性的基本概念 basic concepts of frequency characteristics	（1）定义 （2）频率特性的表示方法 （1）definition （2）representation of frequency characteristics	1
5.2	对数坐标图 log magnitude plots	（1）对数坐标图及其特点 （2）典型环节的对数坐标图 （3）系统对数频率特性的绘制 （4）非最小相位系统的对数坐标图 （1）log magnitude plots and their characteristics （2）log magnitude plots of typical links （3）drawing of logarithmic frequency characteristic of systems （4）log magnitude plots of nonminimum-phase systems	1.5

续表

章节序号 chapter number	章节名称 chapters	知识点 key points	课时 class hour
5.3	极坐标图 polar plots	（1）典型环节的极坐标图 （2）开环系统极坐标图的绘制 （3）非最小相位系统的极坐标图 （4）增加零、极点对极坐标图的影响 （1）polar plots of typical links （2）drawing polar plots of open-loop systems （3）polar plots of nonminimum-phase systems （4）influence of adding zero and pole on polar plots	1.5
5.4	奈奎斯特稳定判据 Nyquist stability criterion	（1）辐角原理 （2）奈奎斯特稳定判据 （3）奈奎斯特稳定判据在伯德图中的应用 （1）principle of argument （2）Nyquist stability criterion （3）application of Nyquist stability criterion in Bode diagram	2.5
5.5	稳定裕度 stability margin	（1）幅值稳定裕度 （2）相位稳定裕度 （3）用稳定裕度确定系统的相对稳定性 （1）magnitude margin （2）phase margin （3）determine the relative stability of the system using stability margin	1.5
5.6	闭环系统的频率特性 frequency characteristics of closed-loop systems	（1）用向量法求闭环频率特性 （2）等幅值轨迹（等M圆）和等相角轨迹（等N圆） （1）using vector method to calculate closed-loop frequency characteristics （2）equi-amplitude trajectory (equi-M circle) and equiphase trajectory (equi-N circle)	1
5.7	闭环系统性能分析 performance analysis of closed-loop systems	（1）利用频率特性分析系统的稳态性能 （2）频域性能指标与时域性能指标的关系 （1）analysis of the steady-state performasnce of the system by using frequency characteristics （2）relationship between frequency-domain performance indices and time-domain performance indices	1

第六章　线性控制系统的设计（design of linear control systems）

章节序号 chapter number	章节名称 chapters	知识点 key points	课时 class hour
6.1	校正装置及其特性 compensation devices and their characteristics	（1）超前校正装置的特性 （2）滞后校正装置的特性 （3）滞后-超前校正装置的特性 （1）characteristics of lead compensator （2）characteristics of lag compensator （3）characteristics of lag-lead compensator	1
6.2	基于伯德图的系统校正 system compensation based on Bode diagram	（1）基于伯德图的相位超前校正 （2）基于伯德图的相位滞后校正 （3）基于伯德图的相位滞后-超前校正 （4）超前-滞后和滞后-超前校正的比较 （1）phase-lead compensation based on Bode diagram （2）phase-lag compensation based on Bode diagram （3）phase lag-lead compensation based on Bode diagram （4）comparison of lead-lag and lag-lead compensation	4
6.3	PID控制器 PID controllers	（1）比例控制器 （2）积分控制器 （3）比例积分（PI）控制器 （4）比例微分（PD）控制器 （5）比例积分微分（PID）控制器 （1）proportional controllers （2）integral controllers （3）proportional-integral (PI) controllers （4）proportional-derivative (PD) controllers （5）proportional-integral-derivative(PID) controllers	1

第七章　线性控制系统的状态空间分析和设计方法（state-space analysis and design methods of linear control systems）

章节序号 chapter number	章节名称 chapters	知识点 key points	课时 class hour
7.1	状态空间模型 state-space model	（1）状态变量和状态变量模型 （2）状态空间表达式的建立方法 （3）传递函数矩阵 （4）状态空间表达式的线性变换 （1）state variables and state variable models （2）method of establishing state-space expressions （3）transfer function matrix （4）linear transformation of state-space expressions	4

续表

章节序号 chapter number	章节名称 chapters	知识点 key points	课时 class hour
7.2	线性控制系统的运动分析 motion analysis of linear control systems	（1）线性连续定常齐次状态方程的解 （2）矩阵指数函数和状态转移矩阵 （3）线性连续定常非齐次状态方程的解 （1）solution of linear continuous time-invariant homogeneous state equations （2）matrix exponential function and state transition matrix （3）solution of the linear continuous time-invariant nonhomogeneous state equations	2
7.3	线性控制系统的能控性和能观测性 controllability and observability of linear control systems	（1）状态能控性的定义及判断方法 （2）状态能观测性的定义及判断方法 （3）对偶原理 （1）definition and decision method of state controllability （2）definition and decision method of state observability （3）duality principle	2
7.4	控制系统的李雅普诺夫稳定性分析 Lyapunov's stability analysis of control systems	（1）李雅普诺夫稳定性的概念 （2）李雅普诺夫第二法 （3）线性连续定常系统的李雅普诺夫稳定性分析 （1）concept of Lyapunov's stability （2）Lyapunov's second method （3）Lyapunov's stability analysis of linear continuous time-invariant systems	2
7.5	状态反馈和状态观测器 state feedback and state observer	（1）状态反馈与极点配 （2）状态观测器设计 （3）带有观测器的状态反馈系统设计 （1）state feedback and pole placement （2）design of state observers （3）design of state feedback system with observers	4

4.5.5 实验环节（Experiments）

序号 number	实验内容 experiment content	知识点 key points	课时 class hour
1	串联校正系统设计 design of series compensation systems	（1）掌握分析原系统特性 （2）根据系统期望值设计超前校正装置的方法 （1）analyze the characteristics of the original systems （2）lead compensator design according to the prescribed performance of the systems	2

续表

序号 number	实验内容 experiment content	知识点 key points	课时 class hour
2	直流电机系统建模与控制 modeling and controlling of DC motor system	（1）掌握软、硬件应用，采用实验测量、参数拟合等方法获得直流电机的模型参数 （2）应用PI、PD控制器设计思想，实现电机的控制 （1）master software and hardware applications, and obtain DC motor model parameters by experimental measurements, parameter fitting, and other methods （2）adopt PI and PD controller design concept to achieve motor control	2
3	模拟直升机起降控制系统设计 design of simulated helicopter take-off and landing control system	（1）掌握LabVIEW软件进行系统设计的方法 （2）应用PID以及状态反馈的控制方法，实现对模拟直升机起降的控制 （1）master the system design method with LabVIEW （2）apply PID and state feedback control method to realize the control of simulated helicopter take-off and landing	4

4.6　"工程图学"教学大纲

课程名称：工程图学

Course Title：Engineering Drawing

先修课程：计算机科学基础与高级程序设计

Prerequisites：Computer Science Fundamentals and Advanced Programming Design

学分：3

Credits：3

4.6.1　课程目的和基本内容（Course Objectives and Basic Contents）

本课程是储能科学与工程专业的一门基础必修课。

工程图样是表达和交流技术思想的重要工具，是工程技术部门的一项重要技术文件。本课程主要研究解决空间几何问题、机械设计表达和阅读工程图样的理论和方法。

This course is a mandatory fundamental course for undergraduates majoring in energy storage science and engineering.

Engineering drawing is an important tool to express and exchange technical ideas, and it is an important technical document of the engineering technology departments. This course mainly studies the theory and method of solving space geometry problems, expressing mechanical design, and reading engineering drawings.

4.6.2 课程基本情况（Basic Information of the Course）

课程名称	工程图学 Engineering Drawing										
开课时间	一年级		二年级		三年级		四年级		总学分	3	总学时	64
	秋	春	秋	春	秋	春	秋	春				
课程定位	储能科学与工程专业本科生储能基础课程群必修课											
授课学时分配	课堂讲授48学时+课外实验16学时											
先修课程	计算机科学基础与高级程序设计											
后续课程	无											
教学方式	课堂教学、讨论、平时作业、上机绘图实践											
考核方式	课程结束笔试成绩占50%，期中考试占25%，计算机绘图作业占10%，平时成绩占15%											
参考教材	[1] 唐克中，郑镁. 画法几何及工程制图.5版.北京,高等教育出版社,2017. [2] 许睦旬，徐凤仙，温伯平. 画法几何及工程制图习题集. 5版. 北京: 高等教育出版社,2017.											
参考资料	[1] 续丹.3D机械制图.北京:机械工业出版社,2008. [2] 续丹.3D机械制图习题集.北京:机械工业出版社,2008. [3] 续丹,许睦旬.三维建模与工程制图.北京:机械工业出版社,2020. [4] 续丹,许睦旬.三维建模与工程制图习题集.北京:机械工业出版社,2020. [5] 续丹,黄胜,等.Solid Edge实践与提高.北京:清华大学出版社,2007. [6] 许睦旬.Inventor 2009 三维机械设计应用基础.北京:高等教育出版社,2009.											

4.6.3 教学目的和基本要求（Teaching Objectives and Basic Requirements）

本课程的主要任务是通过课堂教学、实验教学等环节培养学生的空间思维能力、绘制和阅读机械图样的能力。

课程基本要求：

（1）掌握投影法的基本理论和应用，具有查阅有关标准的能力及绘制和阅读机械图样的基本技能；

（2）能够借助工程图样、软件、模型等载体，完成简单装配体和零件的识读与表达，并能理解机械工程图样所涉及的文字、符号以及图形的含义；

（3）能够掌握一种计算机绘图软件绘制工程图样，逐步具备应用先进绘图设计软件解决工程实际问题的能力。

The main task of this course is to cultivate students' spatial imagination ability, and the ability to draw and read mechanical drawings through classroom teaching and experimental teaching.

Course requirements:

（1）Understand and use the basic theory of projection methods, possess the ability of consulting relevant standards, reading and drawing mechanical drawings.

（2）Be able to complete the recognition and expression of simple assemblies and parts, and understand the meaning of texts, symbols, and graphics involved in mechanical engineering drawings, with the help of engineering drawings, software, models, and other tools.

（3）Be able to master a kind of computer drawing software to create engineering drawings, and gradually have the ability to apply advanced design software to solve practical engineering problems.

4.6.4　课程大纲和知识点（Syllabus and Key Points）

第一章　绪论 (introduction)

章节序号 chapter number	章节名称 chapters	知识点 key points	课时 class hour
1.1	绪论 introduction	（1）了解机械设计表达的过程 （2）明确本课程的地位、性质、任务、发展方向和学习方法 （1）understand the process of mechanical design expression （2）clarify the status, nature, tasks, development direction, and learning methods of this course	1
1.2	制图基本知识 basic knowledge of drawing	（1）熟悉并遵守机械制图国家标准的有关规定 （2）掌握仪器绘图和徒手绘图的方法 （3）掌握常用的几何作图方法 （4）掌握分析和标注平面图形尺寸的方法 （1）be familiar with and abide by the national standards of mechanical drawing （2）master the methods of instrument drawing and freehand drawing （3）master the commonly used geometric drawing methods （4）master the method of analyzing and marking the size of plane graphics	1

第二章 正投影法基础（fundamentals of orthogonal projection）

章节序号 chapter number	章节名称 chapters	知识点 key points	课时 class hour
2.1	投影法概述 overview of projection methods	建立平行投影(正投影和斜投影)的基本概念 establish the basic concept of parallel projection (orthogonal projection and oblique projection)	1
2.2	三视图形成及投影规律 orthographic views formation and projection law	掌握正投影法的基本性质 master the basic properties of the orthogonal projection method	1
2.3	基本几何体（包含回转体）的投影 projection of basic geometries (including cylinder ball of revolution)	掌握基本几何体的投影特性、作图及尺寸标注的基本方法 master the projection characteristics of basic geometries, and the basic methods of drawing and dimensioning	1

第三章 立体的投影分析（analysis of stereoscopic projection）

章节序号 chapter number	章节名称 chapters	知识点 key points	课时 class hour
3.1	点、线、面投影分析 projection of points, lines, planes	（1）掌握点、直线、平面在各种位置下的投影特性和作图基本方法(平面以多边形为主) （2）掌握直线上的点的投影特性以及作图的基本方法 （3）掌握面上点的投影特性，以及面上作点、作线的基本方法	1
3.2	平面立体及回转体（柱、球）面上取点、线 take points and lines on the surface of revolution body (cylinder, ball) and planar objects	（1）master the projection characteristics of points, lines, and planes at various positions and the basic methods of drawing (planes are mainly polygons) （2）master the projection characteristics of points on a straight line and the basic methods of drawing （3）master the projection characteristics of points on the surface, and the basic methods of drawing points and lines on the surface	2

第四章　相交组合体（boolean combination of solids）

章节序号 chapter number	章节名称 chapters	知识点 key points	课时 class hour
4.1	截交、相贯（柱、球） section and intersection （cylinder, ball）	（1）能分析平面与立体相交后所产生的截交线的性质，并掌握作截交线的基本方法 （2）能分析两立体相交后所产生的相贯线的性质，并掌握作相贯线的基本方法 （1）be able to analyze the properties of the section line generated by the intersection of a plane and a solid, and master the basic methods of drawing the section line （2）be able to analyze the properties of the intersection line generated by the intersection of two solids, and master the basic methods of drawing the intersection line	4
4.2	轴测图的概念及正等轴测图画法（自学） concept of axonometric drawing and method of isometric projection drawing (self-study)	建立轴测投影的基本概念，掌握绘制正等轴测图的基本方法（自学） establish the basic concept of axonometric projection and master the basic methods of isometric projection drawing (self-study)	0

第五章　组合体（combination solid）

章节序号 chapter number	章节名称 chapters	知识点 key points	课时 class hour
5.1	组合体的画法 drawing of the combination solid	掌握组合体的画图、读图和尺寸标注的方法 master the methods of drawing, reading, and dimensioning of the combination solid	2
5.2	组合体读图 reading of the combination solid		2
5.3	组合体的尺寸标注 dimensioning of the combination solid		2

第六章 机件形状的表示法（representation of parts）

章节序号 chapter number	章节名称 chapters	知识点 key points	课时 class hour
6.1	视图、剖视图 view and section view	掌握国家标准中规定的各种视图、剖视图的画法 master the drawing method of various views and section views specified in national standards	3
6.2	断面图、简化画法 cross-sectional view and simplified representation	（1）掌握国家标准中规定的断面图的画法 （2）掌握常用的简化画法和其他规定画法 （3）了解第三角画法 （1）master the drawing method of cross-sectional view specified in national standards （2）master the commonly used simplified representation methods and other prescribed drawing methods （3）understand the third angle projection method	3

第七章 零件图（part drawing）

章节序号 chapter number	章节名称 chapters	知识点 key points	课时 class hour
7.1	零件图的作用与内容,零件图视图,零件图上常见工艺结构 function and content of part drawing,view of part drawing,common process structures on part drawing	（1）了解零件图的作用与内容,掌握绘制和阅读典型零件图的基本方法 （2）掌握零件视图选择原则与表达方法 （3）掌握零件上常见工艺结构（螺纹等）的画法和尺寸标注的基本方法 （1）understand the function and content of part drawings, and master the basic methods of drawing and reading tipical part drawings （2）master the selection principle and expression method of part view （3）master the basic methods of drawing and dimensioning of common process structures (thread, etc.) on parts	6
7.2	零件图的尺寸标注与技术要求 dimensions and technical information requirements of part drawing	（1）掌握标注零件图尺寸的基本方法 （2）掌握在零件图上标注表面粗糙度代号、尺寸公差与配合代号的基本方法 （1）master the basic methods of marking the size of the part drawing （2）master the basic methods of marking the surface roughness code, tolerance and fitting code on the part drawing	4

续表

章节序号 chapter number	章节名称 chapters	知识点 key points	课时 class hour
7.3	零件测绘与表达 parts surveying and expression	具有简单零件的测绘表达能力 possess the abilities of surveying and expressing of simple parts	2

第八章　标准件与常用件（standard parts and common parts）

章节序号 chapter number	章节名称 chapters	知识点 key points	课时 class hour
8.1	常用标准件的表示法 representation of commonly-used standard parts	（1）掌握螺纹紧固件（螺栓、螺钉连接）及其连接的规定画法,掌握标准件的规定标记 （2）了解常用件（如平键）的规定画法 （1）master the prescribed drawing method of threaded fasteners (bolts, screw connections) and their connections, and master the prescribed marks of standard parts （2）understand the prescribed drawing methods for common parts (such as flat keys)	3

第九章　装配图（assembly drawing）

章节序号 chapter number	章节名称 chapters	知识点 key points	课时 class hour
9.1	画装配图 drawing assembly drawings	（1）了解装配图的作用与内容 （2）掌握绘制装配图的基本方法 （1）understand the function and content of assembly drawings （2）master the basic methods of drawing assembly drawings	5
9.2	看装配图及拆画零件图 reading assembly drawings and drawing part drawings	掌握阅读简单复杂程度装配图和拆画零件图的基本方法 master the basic methods of reading simple assembly drawings and drawing part drawings	4

第十章　计算机绘图（computer aided drawing）

章节序号 chapter number	章节名称 chapters	知识点 key points	课时 class hour
10.1	计算机绘图 computer aided drawing	能较熟练地使用一种计算机绘图软件绘制工程图样 be able to draw engineering drawings skillfully with a computer aided drawing software	16

4.7 "电路"课程教学大纲

课程名称：电路

Course Title：Electric Circuit

先修课程：高等数学Ⅰ，线性代数与解析几何

Prerequisites：Advanced Mathematics Ⅰ, Linear Algebra and Analytical Geometry

学分：4

Credits：4

4.7.1　课程目的和基本内容（Course Objectives and Basic Contents）

本课程是储能科学与工程专业的一门必修课程。

电路理论将实际电路抽象成电路模型，应用数学方法分析电路在各种工作条件下的行为。通过"电路"课程的学习，树立学生严肃认真的科学作风和理论联系实际的工程观点，培养学生的科学思维能力、分析计算能力和实验研究能力，为后续电类课程的学习奠定必要的电路理论基础。

本课程的基本内容包括电阻电路的分析、一阶二阶动态电路的分析、正弦稳态电路分析、二端口网络。

This course is a mandatory course for undergraduates majoring in energy storage science and engineering.

Circuit theory abstracts the actual circuits into circuit models, and analyzes the behavior of the circuit models under various conditions by mathematical methods. Through the study of "Electric Circuit" course, students are expected to establish serious scientific style and engineering viewpoint of integrating theory with practice, and students' abilities can be cultivated, including scientific thinking, analysis and calculation, and experimental research, to lay the necessary circuit theoretical foundation for the follow-up study of electrical courses.

The basic content of this course includes resistive circuit analysis, first order and second order dynamic circuit analysis, sinusoidal steady-state circuit analysis, and two port networks.

4.7.2 课程基本情况（Basic Information of the Course）

课程名称	电路 Electric Circuit							
开课时间	一年级	二年级	三年级	四年级	总学分	4	总学时	68
	秋　春	秋　春	秋　春	秋　春				
课程定位	储能科学与工程专业本科生储能基础课程群必修课							
授课学时分配	课堂讲授60学时+实验8学时							
先修课程	高等数学Ⅰ,线性代数与解析几何							
后续课程	自动控制理论,现代电子技术,储能系统设计,电力系统分析,嵌入式智能系统,智能电网储能应用技术,电储能系统与并网技术,储能装置设计与开发实验							
教学方式	课堂教学、实验与综述报告、平时作业							
考核方式	课程结束笔试成绩占60%,实验成绩占10%,平时成绩占20%,作业占10%							
参考教材	[1] 邱关源,罗先觉.电路.5版.北京:高等教育出版社,2006.							
参考资料	[1] 陈希有.电路理论教程.北京:高等教育出版社,2013.							

4.7.3 教学目的和基本要求（Teaching Objectives and Basic Requirements）

（1）使学生掌握电路的基本理论、基本知识，为后续相关课程的学习打下坚实的基础；

（2）学生能将相关数学方法应用于电路分析，培养学生的科学思维能力、分析计算能力；

（3）使学生掌握基本实验技能和常见电类实验仪器的使用，获得初步实验研究能力。

（1）Students can master the fundamental theory and knowledge of circuits, and lay a solid foundation for the subsequent related courses.

（2）Students can apply relevant mathematical methods to circuit analysis, and cultivate their ability of scientific thinking, analysis, and calculation.

（3）Students can master the basic experimental skills and the use of common electrical experimental instruments, and obtain the ability of preliminary experimental research.

4.7.4 课程大纲和知识点 (Syllabus and Key Points)

第一章 电路模型和电路定律 (circuit models and circuit laws)

章节序号 chapter number	章节名称 chapters	知识点 key points	课时 class hour
1.1	电路和电路模型 circuits and circuit models	（1）电路概述 （2）电路模型 （1）introduction to circuits （2）circuit models	1
1.2、1.3	电压和电流的参考方向、电功率和能量 reference directions of voltage and current, power and energy	（1）参考方向的引入 （2）电压和电流的参考方向 （3）电功率 （4）能量 （1）introduction of reference directions （2）reference directions of voltage and current （3）power （4）energy	1
1.4—1.7	电路元件 circuit elements	（1）电阻 （2）电压源 （3）电流源 （4）受控电源 （1）resistor （2）voltage source （3）current source （4）dependent source	1
1.8	基尔霍夫定律 Kirchhoff's laws	（1）基尔霍夫电流定律 （2）基尔霍夫电压定律 （1）Kirchhoff's current law （2）Kirchhoff's voltage law	2

第二章 电阻电路的等效变换 (equivalente of resistive circuits)

章节序号 chapter number	章节名称 chapters	知识点 key points	课时 class hour
2.1、2.2	电路的等效变换 equivalent transformation of circuits	（1）等效变换的引入 （2）等效变换的特点 （1）introduction of equivalent transformation （2）characteristics of equivalent transformation	1

续表

章节序号 chapter number	章节名称 chapters	知识点 key points	课时 class hour
2.3、2.4	电阻的等效变换 equivalent transformation of resistances	（1）串联电阻的等效变换 （2）并联电阻的等效变换 （3）Y形和△形连接电阻的等效变换 （1）equivalent transformation of series resistances （2）equivalent transformation of parallel resistances （3）equivalent transformation of resistances with star and delta connections	1
2.5、2.6	电源的等效变换 equivalent transformation of sources	（1）电压源串联和并联的等效变换 （2）电流源串联和并联的等效变换 （3）实际电源的两种模型及其等效变换 （1）equivalence of series and parallel voltage sources （2）equivalence of series and parallel current sources （3）two circuit models for actual sources and their equivalent transformation	1
2.7	输入电阻 input resistance	（1）输入电阻的定义 （2）输入电阻的计算 （1）definition of input resistance （2）calculation of input resistance	1

第三章 电阻电路的一般分析（general analysis of resistive circuits）

章节序号 chapter number	章节名称 chapters	知识点 key points	课时 class hour
3.1、3.2	KCL和KVL的独立方程数 number of independent KCL and KVL equations	（1）电路的图 （2）KCL独立方程数 （3）KVL独立方程数 （1）graphs of circuits （2）number of independent KCL equations （3）number of independent KVL equations	1
3.3—3.5	回路电流法 loop current method	（1）支路电流法 （2）网孔电流法 （3）回路电流法 （1）branch current method （2）mesh current method （3）loop current method	2
3.6	结点电压法 node voltage method	（1）结点电压的引入 （2）结点电压方程的列写	3

续表

章节序号 chapter number	章节名称 chapters	知识点 key points	课时 class hour
3.6	结点电压法 node voltage method	（1）introduction of node voltage （2）write out node voltage equations	3

第四章 电路定理（circuit theorems）

章节序号 chapter number	章节名称 chapters	知识点 key points	课时 class hour
4.1、4.2	叠加定理、替代定理 superposition theorem and substitution theorem	（1）叠加定理 （2）替代定理 （1）superposition theorem （2）substitution theorem	1
4.3	戴维南定理和诺顿定理 Thevenin's theorem and Norton's theorem	（1）戴维南定理 （2）诺顿定理 （1）Thevenin's theorem （2）Norton's theorem	3
4.4	最大功率传输定理 maximum power transfer theorem	（1）最大功率传输定理的证明 （2）通过戴维南等效电路计算最大功率 （1）proof of maximum power transfer theorem （2）calculate the maximum power through Thevenin's equivalent circuit	1
4.5	特勒根定理 Tellegen's theorems	（1）特勒根定理1 （2）特勒根定理2 （1）Tellegen's theorem 1 （2）Tellegen's theorem 2	1
4.6、4.7	互易定理、对偶原理 reciprocity theorem, duality principle	（1）互易定理 （2）对偶原理 （1）reciprocity theorem （2）duality principle	1

第五章 含有运算放大器的电阻电路（resistive circuits with operational amplifiers）

章节序号 chapter number	章节名称 chapters	知识点 key points	课时 class hour
5.1	运算放大器的电路模型 circuit model of operational amplifiers	（1）运算放大器简介 （2）运算放大器的电路模型 （1）introduction to operational amplifiers （2）circuit model of operational amplifiers	1

续表

章节序号 chapter number	章节名称 chapters	知识点 key points	课时 class hour
5.2、5.3	含有理想运算放大器电路的分析 analysis of circuits with ideal operational amplifiers	（1）理想运算放大器的特性 （2）含有理想运算放大器电路的分析 （1）characteristics of ideal operational amplifiers （2）analysis of circuits with ideal operational amplifiers	1

第六章　储能元件（energy storage elements）

章节序号 chapter number	章节名称 chapters	知识点 key points	课时 class hour
6.1	电容 capacitors	（1）电容 （2）电容的串联和并联 （1）capacitors （2）series and parallel of capacitors	1
6.2	电感 inductors	（1）电感 （2）电感的串联和并联 （1）inductors （2）series and parallel of inductors	1

第七章　一阶电路和二阶电路的时域分析（time domain analysis of first order and second-order circuits）

章节序号 chapter number	章节名称 chapters	知识点 key points	课时 class hour
7.1	动态电路的方程及其初始条件 equations and initial conditions of dynamic circuits	（1）动态电路微分方程的列写 （2）电容电压和电感电流初值的确定 （1）write out the differential equations of dynamic circuits （2）determination of the initial values of capacitor voltage and inductor current	1
7.2—7.4	一阶电路 first order circuit	（1）一阶电路的零输入响应 （2）一阶电路的零状态响应 （3）一阶电路的全响应 （1）zero input response of first order circuit （2）zero state response of first order circuit （3）full response of first order circuit	4

续表

章节序号 chapter number	章节名称 chapters	知识点 key points	课时 class hour
7.5、7.6	二阶电路 second order circuit	（1）二阶电路的微分方程及其求解 （2）二阶电路的三种工作状态（过阻尼、临界阻尼和欠阻尼） （1）differential equation of second order circuit and its solution （2）three operating states of second order circuit (over damping, critical damping, and under damping)	2
7.7	一阶电路和二阶电路的阶跃响应 step response of first order and second order circuits	（1）阶跃函数 （2）阶跃响应 （1）step function （2）step response	1
7.8	一阶电路和二阶电路的冲激响应 impulse response of first order and second order circuits	（1）冲激函数 （2）冲激响应 （1）impulse function （2）impulse response	1

第八章 相量法（phasor method）

章节序号 chapter number	章节名称 chapters	知识点 key points	课时 class hour
8.1—8.3	相量法基础 fundamentals of phasor method	（1）正弦量 （2）复数 （3）相量的引入 （1）sinusoidal quantity （2）complex number （3）introduction of phasor	1
8.4	电路定律的相量形式 phasor representation of circuit laws	（1）相量的运算 （2）基尔霍夫定律的相量形式 （1）phasor operation （2）phasor representation of Kirchhoff's laws	1

第九章　正弦稳态电路的分析（analysis of sinusoidal steady-state circuits）

章节序号 chapter number	章节名称 chapters	知识点 key points	课时 class hour
9.1	阻抗和导纳 impedance and admittance	（1）阻抗 （2）导纳 （1）impedance （2）admittance	1
9.2	电路的相量图 phasor diagram of circuits	（1）相量图的绘制方法 （2）借助相量图分析电路 （1）drawing method of phasor diagram （2）analysis of circuits by phasor diagram	1
9.3	正弦稳态电路的分析 analysis of sinusoidal steady-state circuits	（1）结点电压法求解正弦稳态电路 （2）正弦稳态电路的戴维南等效 （1）nodal voltage method for solving sinusoidal steady-state circuits （2）Thevenin equivalence of sinusoidal steady-state circuits	1
9.4—9.6	正弦稳态电路的功率 power of sinusoidal steady-state circuits	（1）瞬时功率 （2）平均功率（有功功率） （3）无功功率 （4）视在功率 （5）复功率 （6）功率因数 （7）功率因数的提高 （8）最大功率传输 （1）instantaneous power （2）average power (active power) （3）reactive power （4）apparent power （5）complex power （6）power factor （7）improve power factor （8）maximum power transfer	2

第十章 含有耦合电感的电路（circuits with coupled inductors）

章节序号 chapter number	章节名称 chapters	知识点 key points	课时 class hour
10.1	互感 mutual inductance	（1）互感的定义 （2）同名端 （1）definition of mutual inductance （2）dot convention	1
10.2、10.3	含有耦合电感电路的计算 calculation of circuits with coupled inductors	（1）互感电压 （2）互感的去耦等效 （1）mutual inductance voltage （2）decoupling equivalence of mutual inductance	1
10.4	变压器原理 principle of transformers	（1）变压器 （2）反映阻抗 （1）transformers （2）reflection impedance	1
10.5	理想变压器 ideal transformers	（1）理想变压器的特性 （2）通过理想变压器实现阻抗变换 （1）characteristics of ideal transformers （2）impedance transformation by ideal transformers	1

第十一章 电路的频率响应（frequency response of circuits）

章节序号 chapter number	章节名称 chapters	知识点 key points	课时 class hour
11.1	网络函数 network functions	（1）网络函数的定义 （2）网络函数的幅频特性和相频特性 （1）definition of network functions （2）magnitude-frequency and phase-frequency characteristics of network functions	1
11.2—11.4	谐振 resonance	（1）谐振定义 （2）RLC串联谐振 （3）RLC并联谐振 （1）definition of resonance （2）RLC series resonance （3）RLC parallel resonance	2
11.5	滤波器简介 introduction to filters	（1）滤波器的类型 （2）滤波器的幅频特性 （1）type of filters （2）amplitude frequency characteristics of filters	1

第十二章　三相电路（three-phase circuits）

章节序号 chapter number	章节名称 chapters	知识点 key points	课时 class hour
12.1、12.2	三相电路 three-phase circuits	（1）三相电路简介 （2）三相电路的线电压（电流）和相电压（电流）的关系 （1）introduction to three-phase circuits （2）relationship between line voltage (current) and phase voltage (current) of three-phase circuits	1
12.3	对称三相电路的计算 calculation of symmetrical three-phase circuits	（1）对称三相电路的定义 （2）对称三相电路的计算 （1）definition of symmetrical three-phase circuits （2）calculation of symmetrical three-phase circuits	1
12.4	不对称三相电路的概念 concepts of asymmetrical three-phase circuits	（1）不对称三相电路的定义 （2）不对称三相电路的计算 （1）definition of asymmetrical three-phase circuits （2）calculation of asymmetrical three-phase circuits	1
12.5	三相电路的功率 power of three-phase circuits	（1）对称三相电路的瞬时功率 （2）二瓦计法 （1）instantaneous power of symmetrical three-phase circuits （2）two-wattmeter method	1

第十三章　非正弦周期电流电路（non-sinusoidal periodic current circuits）

章节序号 chapter number	章节名称 chapters	知识点 key points	课时 class hour
13.1— 13.3	非正弦周期信号 non-sinusoidal periodic signals	（1）非正弦周期函数分解为傅里叶级数 （2）非正弦周期信号的有效值和平均功率 （1）decomposition of non-sinusoidal periodic functions into Fourier series （2）RMS value and average power of non-sinusoidal periodic signals	1
13.4	非正弦周期电流电路的计算 calculation of non-sinusoidal periodic current circuits	（1）叠加定理回顾 （2）根据叠加定理计算非正弦周期电流电路 （1）review of superposition theorem （2）calculation of non-sinusoidal periodic current circuit based on superposition theorem	1

第十四章　二端口网络（two-port network）

章节序号 chapter number	章节名称 chapters	知识点 key points	课时 class hour
14.1、14.2	二端口网络及其方程和参数 two-port network and its equations and parameters	（1）二端口网络的定义 （2）二端口网络Z、Y、H、T参数的定义及其求解 （1）definition of two-port network （2）Z、Y、H、T parameters of two-port network and their calculation	1
14.3、14.4	二端口网络的等效电路、二端口网络的连接 equivalent circuits and connections of two-port networks	（1）二端口网络的等效电路 （2）二端口网络的连接 （1）equivalent circuits of two-port network （2）connections of two-port networks	1

第十五章　非线性电路（nonlinear circuits）

章节序号 chapter number	章节名称 chapters	知识点 key points	课时 class hour
15.1、15.2	非线性电路及其方程 nonlinear circuits and their equations	（1）非线性电路简介 （2）非线性电阻电路的方程 （1）introduction to nonlinear circuits （2）equations of nonlinear resistive circuits	1
15.3	非线性电路的小信号分析 small-signal analysis of nonlinear circuits	（1）小信号分析的原理 （2）非线性电路小信号分析的步骤 （1）principle of small-signal analysis （2）steps of small-signal analysis for nonlinear circuits	1

4.7.5　实验环节（Experiments）

序号 number	实验内容 experiment content	知识点 key points	课时 class hour
1	示波器的使用 use of oscilloscope	（1）观察并记录示波器内置标准方波信号的参数 （2）使用示波器测量正弦信号 （3）测量相位差 （4）在Multisim软件中仿真电路,学习用虚拟示波器观察并测量电源电压和电阻电压相位差的方法	2

续表

序号 number	实验内容 experiment content	知识点 key points	课时 class hour
1	示波器的使用 use of oscilloscope	（1）observe and record the parameters of the square wave signal in the oscilloscope （2）use oscilloscope to measure sine signal （3）measure the phase difference （4）simulate in Multisim software, and study the method of observing and measuring the phase difference of the source voltage and resistance voltage with virtual oscilloscope	2
2	直流电阻电路 DC resistance circuit	（1）基尔霍夫定律实验演示与分析 （2）电路元件伏安特性测量 （3）戴维南等效电路测量 （1）experimental demonstration and analysis of Kirchhoff's laws （2）measurement of Volt-Ampere characteristics of circuit elements （3）measurement of Thevenin's equivalent circuit	2
3	线性动态电路 linear dynamic circuits	（1）一阶RC电路 （2）二阶RLC电路 （3）在Multisim进行动态电路的暂态响应仿真 （1）first order RC circuit. （2）second order RLC circuit （3）simulate the transient response of the dynamic circuits in Multisim	2
4	滤波器和谐振电路 filter and resonant circuit	（1）RC高通滤波器实验 （2）基于运算放大器的有源滤波器实验 （3）RLC带通滤波器和谐振实验 （4）Multisim交流电路扫频仿真 （1）RC high pass filter experiment （2）experiment of active power filter based on operational amplifier （3）RLC bandpass filter and resonance experiment （4）Multisim AC circuit sweep simulation	2

4.8 "现代电子技术"课程教学大纲

课程名称：现代电子技术

Course Title：Modern Electronic Technology

先修课程：高等数学Ⅰ，电路

Prerequisites：Advanced Mathematics Ⅰ, Electric Circuit

学分：3

Credits：3

4.8.1 课程目的和基本内容（Course Objectives and Basic Contents）

本课程是储能科学与工程专业的一门基础必修课。

本课程的任务是使学生获得模拟电子技术以及数字电路技术方面的基本理论、基本知识和基本技能，培养学生分析问题和解决问题的能力，为以后深入学习电子技术某些领域打好基础。现代电子技术包括模拟电子技术和数字电子技术基础，主要内容包括半导体二极管及其应用、晶体管及放大电路基础、场效应管及其放大电路、集成运算放大器、反馈和负反馈放大电路、信号运算电路、信号检测与处理电路、信号发生器、功率放大器、直流稳压电源、数字逻辑基础、集成逻辑门电路、组合逻辑电路的分析与设计、触发器、时序逻辑电路、脉冲的产生与整形、数模和模数转换等。

现代电子技术是一门实践性很强的课程，开设的实验以模拟电子技术和数字电子技术的知识为背景。通过实验学习电子电路设计的基本方法、电子电路检测基本知识（包括电子电路测量技术的基本知识，常用电子测量仪器的正确使用，电子电路调试与故障检测的基本方法），并培养正确记录与处理数据、合理表达与报告实验结果的能力。

This course is a mandatory fundamental course for undergraduates majoring in energy storage science and engineering.

This course enables students to acquire basic theories, basic knowledge, and basic skills in analog electronic technology and digital circuit technology. It develops students' ability to analyze and solve problems and lay a solid foundation for future study. Modern electronic technology includes two parts which are analog electronic technology and digital electronic technology, including semiconductor diodes and their applications, transistors and amplifier circuits, field-effect transistors and their amplifier circuits, integrated operational amplifiers, feedback and negative feedback amplifier circuits, and signal operations analysis and design of circuits, signal detection and processing circuits, signal generators, power amplifiers, DC stabilized power supplies, digital logic foundations, integrated logic-gate circuits, combinational logic circuits, flip-flops, sequential logic circuits, pulse generation and shaping, and digital-to-analog and analog-to-digital conversions etc.

Modern electronic technology is a course with very strong practicality. The experiments are based on the knowledge of analog electronic technology and digital electronic technology. Through experiments, students can learn basic methods to design electronic circuits and knowledge to detect electronic circuits, including electronic circuit measurement technology, the correct use of common electronic measuring instruments, and the basic concepts of electronic circuit debugging and fault detection. Students can acquire abilities such as correct recording and processing of test data and reasonable expressing and reporting experimental results.

4.8.2　课程基本情况（Basic Information of the Course）

课程名称	现代电子技术 Modern Electronic Technology									
开课时间	一年级		二年级		三年级		四年级		总学分	3	总学时	56
	秋	春	秋	春	秋	春	秋	春				
课程定位	储能科学与工程专业本科生储能基础课程群（C模块）必修课									
授课学时分配	课堂讲授40学时+实验16学时									
先修课程	高等数学Ⅰ,电路									
后续课程	自动控制理论,测控技术									
教学方式	课堂教学、实验与综述报告、平时作业									
考核方式	课程结束笔试成绩占70%,实验报告占20%,平时成绩占10%									
参考教材	[1] 赵进全,杨拴科. 模拟电子技术基础. 3版.北京:高等教育出版社,2019. [2] 赵进全,张克农. 数字电子技术基础. 3版.北京:高等教育出版社,2020.									
参考资料	[1] 赵进全,杨拴科. 模拟电子技术基础学习指导与解题指南. 3版.北京:高等教育出版社,2021. [2] 赵进全,张克农. 数字电子技术基础学习指导与解题指南. 3版.北京:高等教育出版社,2021.									

4.8.3　教学目的和基本要求（Teaching Objectives and Basic Requirements）

通过本课程的学习,使学生获得现代电子技术方面的基本理论、基本知识和基本技能,为后续相关课程的学习打下一定的基础。

课程基本要求:

（1）了解半导体器件的基本特性和工作原理;

（2）了解放大电路的组成、工作原理、分析方法及放大电路的基本应用;

（3）了解集成器件的应用;

（4）了解数字电路的一些基本知识;

（5）了解常用集成逻辑器件及其应用;

（6）了解逻辑电路的分析及设计方法；

（7）了解数模和模数转换的基本原理；

（8）了解现代电子技术发展特点及发展方向。

Through the study of this course, students can obtain the basic theory, knowledge, and skills of modern electronic technology, and lay the foundation for the subsequent study of related courses.

Basic requirements:

（1）Understand the basic characteristics and working principles of semiconductor device.

（2）Understand the composition, working principle, analysis method, and basic application of the amplifier circuit.

（3）Understand the application of integrated devices.

（4）Understand basic knowledge of digital circuit.

（5）Understand the commonly used integrated logic devices and their applications.

（6）Understand the analysis and design methods of logic circuits.

（7）Understand the basic principles of digital-to-analog and analog-to-digital conversion.

（8）Understand the characteristic and development of modern electronic technology.

4.8.4　课程大纲和知识点（Syllabus and Key Points）

第一章　半导体二极管及其应用（semiconductor diode and its application）

章节序号 chapter number	章节名称 chapters	知识点 key points	课时 class hour
1.1	半导体二极管 semiconductor diode	（1）二极管的结构和类型 （2）二极管的伏安特性 （1）structure and type of diode （2）Volt-Ampere characteristic of diode	1
1.2	半导体二极管（包括稳压二极管）的应用 application of semiconductor diode (including zener diode)	（1）半导体二极管的应用 （2）稳压二极管的应用 （1）application of semiconductor diode （2）application of zener diode	1

第二章 晶体管及其放大电路基础（transistor and its amplifying circuit）

章节序号 chapter number	章节名称 chapters	知识点 key points	课时 class hour
2.1	晶体管的工作原理 working principle of transistor	（1）晶体管的结构 （2）晶体管的工作原理 （3）晶体管的伏安特性 （1）structure of transistor （2）working principle of transistor （3）the Volt-Ampere characteristic of transistor	2
2.2	晶体管放大电路及其分析 transistor amplifier circuit and its analysis	（1）共射极放大电路 （2）共集电极放大电路 （1）common emitter amplifier circuit （2）common collector amplifier circuit	2
2.3	多级放大电路 multi-stage amplifier circuits	（1）多级放大电路的组成 （2）多级放大电路的分析 （1）composition of multi-stage amplifier circuits （2）analysis of multi-stage amplifier circuits	1

第三章 场效应管及其放大电路（field effect transistor and its amplifying circuit）

章节序号 chapter number	章节名称 chapters	知识点 key points	课时 class hour
3.1	场效应管 field effect transistor	（1）场效应管的结构和类型 （2）场效应管的工作原理 （1）structure and type of field effect transistor （2）working principle of field effect transistor	1.5
3.2	场效应管放大电路及其分析 amplifier circuit of field effect transistor and its analysis	（1）场效应管放大电路 （2）场效应管放大电路分析 （1）amplifier circuit of field effect transistor （2）analysis of field effect transistor amplifier circuit	1.5

第四章 负反馈放大电路（negative feedback amplifier circuit）

章节序号 chapter number	章节名称 chapters	知识点 key points	课时 class hour
4.1	负反馈的基本概念 basic concept of negative feedback	（1）负反馈的类型 （2）负反馈的判断 （1）types of negative feedback （2）judgement of negative feedback	1.5

续表

章节序号 chapter number	章节名称 chapters	知识点 key points	课时 class hour
4.2	负反馈对放大电路性能的影响 effect of negative feedback on the performance of the amplifier circuit	（1）稳定放大倍数 （2）减小非线性失真 （3）影响输入输出电阻 （1）stable magnification （2）reduce nonlinear distortion （3）effect input and output resistance	1
4.3	负反馈放大电路的近似计算 approximate calculation of negative feedback amplifier circuit	（1）集成元件电路的计算 （2）分立元件电路的计算 （1）calculation of integrated component circuit （2）calculation of discrete component circuit	1

第五章 运算电路（arithmetic circuit）

章节序号 chapter number	章节名称 chapters	知识点 key points	课时 class hour
5.1	运算放大器组成的运算电路 operational circuit composed of operational amplifiers	（1）加法电路 （2）减法电路 （3）积分电路 （4）微分电路 （5）除法电路 （1）adding circuit （2）subtraction circuit （3）integral circuit （4）differential circuit （5）division circuit	1.5

第六章 信号处理电路（signal processing circuit）

章节序号 chapter number	章节名称 chapters	知识点 key points	课时 class hour
6.1	有源滤波器 active filter	（1）有源滤波器的基本概念 （2）一阶有源滤波器 （3）二阶有源滤波器 （1）basic concept of active filter （2）first-order active filter （3）second-order active filter	1.5

续表

章节序号 chapter number	章节名称 chapters	知识点 key points	课时 class hour
6.2	电压比较器 voltage comparator	（1）单门限电压比较器 （2）多门限电压比较器 （1）single threshold voltage comparator （2）multi-threshold voltage comparator	1.5

第七章　信号产生电路（signal generating circuit）

章节序号 chapter number	章节名称 chapters	知识点 key points	课时 class hour
7.1	正弦波发生器 sine wave generator	（1）正弦波发生器的原理及组成 （2）RC型正弦波发生器 （3）LC型正弦波发生器 （1）principle and composition of sine wave generator （2）RC type sine wave generator （3）LC type sine wave generator	1.5
7.2	非正弦波发生器 non-sine wave generator	（1）方波发生器 （2）方波—三角波发生器 （1）square-wave generator （2）square-wave-triangle-wave generator	1.5

第八章　功率放大电路（power amplifier circuit）

章节序号 chapter number	章节名称 chapters	知识点 key points	课时 class hour
8.1	功率放大电路的特点及分类 features and classification of power amplifier circuit	（1）功率放大电路的特点 （2）功率放大电路的分类 （1）features of power amplifier circuit （2）classification of power amplifier circuit	0.5
8.2	互补推挽功率放大电路 complementary push-pull power amplifier circuit	（1）乙类互补推挽功率放大电路 （2）甲、乙类互补推挽功率放大电路 （1）class B complementary push-pull power amplifier circuit （2）class A and B complementary push-pull power amplifier circuit	1

第九章 直流稳压电源（DC power supply）

章节序号 chapter number	章节名称 chapters	知识点 key points	课时 class hour
9.1	整流滤波电路 rectifier filter circuit	（1）单相桥式整流电路 （2）单相桥式整流电容滤波电路 （1）single-phase bridge rectifier circuit （2）single-phase bridge rectifier capacitor filter circuit	1
9.2	稳压电路 voltage stabilizing circuit	（1）串联反馈型线性稳压电路 （2）三端集成稳压器 （1）series feedback linear voltage stabilizing circuit （2）three-terminal integrated voltage regulators	0.5

第十章 数字逻辑基础（foundations of digital logic）

章节序号 chapter number	章节名称 chapters	知识点 key points	课时 class hour
10.1	数制与数码 number system and digital	（1）数制 （2）数码 （1）number system （2）digital	0.5
10.2	基本逻辑运算 basic logic operations	（1）基本逻辑运算 （2）复合逻辑运算 （1）basic logic operations （2）compound logic operations	1
10.3	逻辑函数及其化简 logical function and its simplification	（1）逻辑函数的代数化简 （2）逻辑函数的卡诺图化简 （1）algebraic simplification of logic function （2）Karnaugh map simplification of logic function	1.5

第十一章 集成逻辑门电路（integrated logic gate）

章节序号 chapter number	章节名称 chapters	知识点 key points	课时 class hour
11.1	TTL门电路 TTL gate circuit	（1）TTL门电路的工作原理 （2）TTL门电路的外特性 （1）working principle of TTL gate circuit （2）external characteristic of TTL gate circuit	0.5

续表

章节序号 chapter number	章节名称 chapters	知识点 key points	课时 class hour
11.2	CMOS门电路 CMOS gate circuit	（1）COMS门电路的工作原理 （2）CMOS门电路的外特性 （1）working principle of COMS gate circuit （2）external characteristic of CMOS gate circuit	0.5

第十二章　组合逻辑电路的分析与设计（analysis and design of combinational logic circuit）

章节序号 chapter number	章节名称 chapters	知识点 key points	课时 class hour
12.1	基于门电路的组合逻辑电路分析与设计 analysis and design of combinational logic circuit based on gate circuit	（1）基于门电路的组合逻辑电路分析 （2）基于门电路的组合逻辑电路设计 （1）analysis of combinational logic circuit based on gate circuit （2）design of combinational logic circuit based on gate circuit	1
12.2	常用的中规模组合逻辑器件 commonly used medium-scale combinational logic devices	常用的中规模组合逻辑器件介绍 introduction to commonly used medium-scale combinational logic devices	1.5
12.3	基于MSI的组合逻辑电路分析与设计 analysis and design of combinational logic circuit based on MSI	（1）基于MSI的组合逻辑电路分析 （2）基于MSI的组合逻辑电路设计 （1）analysis of combinational logic circuit based on MSI （2）design of combinational logic circuit based on MSI	1.5

第十三章　触发器（trigger）

章节序号 chapter number	章节名称 chapters	知识点 key points	课时 class hour
13.1	触发器及其功能 trigger and its function	（1）D触发器 （2）JK触发器 （3）T触发器 （1）D trigger （2）JK trigger （3）T trigger	1

第十四章 时序逻辑电路（sequential logic circuit）

章节序号 chapter number	章节名称 chapters	知识点 key points	课时 class hour
14.1	基于触发器的时序逻辑电路分析与设计 analysis and design of sequential logic circuit based on trigger	（1）基于触发器的时序逻辑电路分析 （2）基于触发器的时序逻辑电路设计 （1）analysis of sequential logic circuit based on trigger （2）design of sequential logic circuit based on trigger	1
14.2	中规模时序逻辑器件 medium-scale sequential logic devices	常用的中规模时序逻辑器件介绍 introduction to commonly used medium-scale sequential logic devices	1.5
14.3	基于MSI的时序逻辑电路分析与设计 analysis and design of sequential logic circuits based on MSI	（1）基于MSI的时序逻辑电路分析 （2）基于MSI的时序逻辑电路设计 （1）analysis of sequential logic circuit based on MSI （2）design of sequential logic circuit based on MSI	1

第十五章 脉冲的产生与整形（pulse generation and shaping）

章节序号 chapter number	章节名称 chapters	知识点 key points	课时 class hour
15.1	555定时器 555 timer	（1）555定时器的电路结构 （2）555定时器的工作原理 （1）circuit structure of 555 timer （2）working principle of 555 timer	0.5
15.2	555定时器的应用 application of 555 timer	（1）施密特触发器 （2）单稳态电路 （3）多谐振荡器 （1）schmidt trigger （2）monostable circuit （3）multivibrator	1

第十六章 数模和模数转换（digital-to-analog and analog-to-digital conversion）

章节序号 chapter number	章节名称 chapters	知识点 key points	课时 class hour
16.1	数模转换 digital-to-analog conversion	（1）数模转换原理 （2）权电阻网路DAC （1）principle of digital-to-analog conversion （2）right resistance network DAC	1

续表

章节序号 chapter number	章节名称 chapters	知识点 key points	课时 class hour
16.2	模数转换 analog-to-digital conversion	（1）模数转换原理 （2）并行比较型DAC （3）逐次渐进型DAC （1）principle of analog-to-digital conversion （2）parallel comparison type DAC （3）progressive DAC	1

4.8.5　实验环节（Experiments）

序号 number	实验内容 experiment content	知识点 key points	课时 class hour
1	单管放大电路 single tube amplifier circuit	（1）搭建电路 （2）调节静态工作点 （3）测试电路性能指标 （4）观测静态工作点和信号频率对输出的影响 （1）build the circuit （2）adjust the static working point （3）test circuit performance （4）observe the effect of static operating point and signal frequency on the output signals	4
2	方波—三角波发生器 square-wave-triangle-wave generator	（1）搭建电路 （2）调节电路 （3）测试电路性能指标 （1）build the circuit （2）regulate the circuit （3）test the circuit performance	6
3	简易数字钟 simple digital clock	（1）搭建电路 （2）调试测试 （1）build the circuit （2）circuit debugging and testing	6

4.9　"储能原理 I"教学大纲

课程名称：储能原理 I

Course Title：Energy Storage Principle I

先修课程：高等数学Ⅰ，储能化学基础（含实验）

Prerequisites：Advanced Mathematics Ⅰ, General Chemistry for Energy Storage（with Experiments）

学分：3

Credits：3

4.9.1 课程目的和基本内容（Course Objectives and Basic Contents）

本课程涵盖了储能科学与技术的一些基本知识，兼顾关键科学理论与实际工程应用，深入浅出地介绍各种储能技术的工作原理和特性，力争反映我国储能领域的新进展，为学生掌握储能技术的基本原理和储能领域的最新进展打下坚实的基础。

储能原理主要介绍各种储能技术的新进展、应用范围、产业现状、技术经济性等，同时对储能技术在电网、交通、新能源等领域的应用进行了详尽分析。

This course covers some basic knowledge of energy storage science and technology, considering key scientific theories and practical engineering applications. It introduces the working principles and characteristics of various energy storage technologies in depth and tries to reflect the new progress in the field of energy storage in China. It lays a solid foundation for students to master the basic principles of energy storage technology and the latest progress in the field of energy storage.

This course mainly introduces the new progress, application scope, industrial development status, and technical economy of various energy storage technologies etc. Meanwhile, the application of energy storage technologies is analyzed in detail, including the power grid, transportation, alternative energy, and other fields.

4.9.2 课程基本情况（Basic Information of the Course）

课程名称	储能原理 I Energy Storage Principle Ⅰ											
开课时间	一年级		二年级		三年级		四年级		总学分	3	总学时	48
	秋	春	秋	春	秋	春	秋	春				
课程定位	储能科学与工程专业本科生储能基础课程群必修课											
授课学时分配	课堂讲授48学时											
先修课程	高等数学Ⅰ,储能化学基础（含实验）											
后续课程	可再生能源及其发电技术,储能材料工程,储能系统设计,储能系统检测与估计,大型储能工程导论											
教学方式	课堂教学,讨论,平时作业											
考核方式	课程结束笔试成绩占70%,平时成绩占30%											

续表

参考教材	[1] 丁玉龙,来小康,陈海生. 储能技术及应用. 北京:化学工业出版社:2018. [2] Barnes F B, Levine J G.大规模储能系统. 肖曦,聂赞相,译.北京:机械工业出版社,2018. [3] Brunet Y;储能技术及应用.唐西胜,徐鲁宁,周龙,等,译.北京:机械工业出版社,2018.
参考资料	无

4.9.3 教学目的和基本要求（Teaching Objectives and Basic Requirements）

（1）培养学生树立正确的人生观、价值观、社会观和科学观，有较高的思想道德、社会责任感、文化素养和专业素质，富有求实创新的意识；

（2）使学生扎实掌握本专业基础理论知识、专业技能和应用技术；

（3）使学生具备一定的学习能力、实践能力、研究能力和新技术开发能力；

（4）使学生接受科学的训练，具有综合运用基础理论、技术、方法及计算模拟解决实际问题的能力，具有良好的科学素养、系统思维能力，具有良好的外语阅读、交流与写作能力，具有良好的国际化视野；

（5）使学生具备在能源与动力工程、电气工程及其自动化、材料物理、材料化学及相关学科进一步深造的基础，或满足教学、科研、技术开发以及管理等方面工作的要求。

（1）Cultivate students to establish a correct outlook on life, values, social view, and scientific view. Students are expected to have a high ideological and moral, social responsibility, cultural literacy, professional quality, and full of realistic and innovative consciousness.

（2）Master the basic theoretical knowledge, professional skills, and applied technology of energy storage.

（3）Through this course, train students to have certain learning abilities, practical abilities, research abilities, and abilities to develop new technology.

（4）Train students scientifically, enable them own the ability to solve practical problems by comprehensively applying basic theory, technology, method, and computer simulation technique. Meanwhile, train them to possess good scientific literacy, systematic thinking ability, foreign language ability, and international vision.

（5）Enable students to have the ability of constantly studying in energy and power engineering, electrical engineering, material physics and chemistry, and related disciplines, so as to meet the requirements of teaching, research, technology development, and management.

4.9.4 课程大纲和知识点（Syllabus and Key Points）

第一章 绪论（introduction）

章节序号 chapter number	章节名称 chapters	知识点 key points	课时 class hour
1.1	储能技术简介 introduction of energy storage technology	（1）储能技术的重要性与主要功能 （2）储能技术的多样性 （1）importance and main functions of energy storage technology （2）diversity of energy storage technology	2
1.2	储能技术分类与应用 classification and application of energy storage technology	（1）储能技术的分类与发展程度 （2）储能技术应用现状和市场预测 （3）储能技术的研究情况 （1）classification and development of energy storage technology （2）application status and market forecast of energy storage technology （3）research status of energy storage technology	2

第二章 锂离子电池技术（lithium-ion battery technology）

章节序号 chapter number	章节名称 chapters	知识点 key points	课时 class hour
2.1	锂离子电池介绍 introduction of lithium-ion batteries	（1）锂离子电池的发展历史概述和基本原理 （2）锂离子电池的功率和能量应用范围 （1）overview of the development history and basic principles of lithium-ion batteries （2）range of power and energy applications of lithium-ion batteries	3
2.2	锂离子电池关键材料 key materials for lithium-ion batteries	（1）锂离子电池关键材料概述 （2）锂离子电池关键材料的发展现状 （1）overview of key materials for lithium-ion batteries （2）development status of key materials for lithium-ion batteries	3
2.3	锂离子电池的发展现状 development status of lithium-ion batteries	（1）能量型锂离子电池的技术发展和应用现状 （2）动力型锂离子电池技术的发展现状 （3）储能型锂离子电池的发展现状	3

续表

章节序号 chapter number	章节名称 chapters	知识点 key points	课时 class hour
2.3	锂离子电池的发展现状 development status of lithium-ion batteries	（1）technology development and application status of energy-type lithium-ion batteries （2）development status of power-type lithium-ion battery technology （3）development status of energy storage lithium-ion batteries	3
2.4	锂离子电池的技术指标及未来发展线路图 technical specifications and future development roadmap of lithium-ion battery	（1）锂离子电池的技术指标 （2）锂离子电池的未来发展线路图 （1）technical specifications of lithium-ion battery （2）future development roadmap of lithium-ion battery	3

第三章　液流电池技术（flow battery technology）

章节序号 chapter number	章节名称 chapters	知识点 key points	课时 class hour
3.1	液流电池简介 introduction of flow batteries	（1）液流电池的基本原理和发展历史概述 （2）几种典型的液流电池体系 （1）overview of the basic principle and development history of flow batteries （2）several typical flow battery systems	1
3.2	液流电池的关键材料 key materials for flow batteries	（1）液流电池的关键材料 （2）液流电池的效率与影响因素分析 （1）key materials for flow batteries （2）analysis of efficiency and influencing factors of flow batteries	2
3.3	液流电池的经济和技术指标及未来发展展望 economic and technical index and future development prospect of flow batteries	（1）液流电池的经济和技术指标 （2）液流电池的未来发展展望 （1）economic and technical index of flow batteries （2）future development prospect of flow batteries	1

第四章 全钒液流电池技术（full vanadium flow battery technology）

章节序号 chapter number	章节名称 chapters	知识点 key points	课时 class hour
4.1	全钒液流电池概述 overview of all vanadium flow batteries	（1）全钒液流电池关键材料 （2）全钒液流电池电堆、系统管理与控制系统 （1）key materials for all vanadium flow batteries （2）reactor, system management, and control system of all vanadium liquid flow batteries	2
4.2	全钒液流电池的应用及前景分析 application and prospect analysis of all vanadium flow batteries	（1）全钒液流电池的应用 （2）全钒液流电池的前景分析 （1）application of all vanadium flow batteries （2）prospect analysis of all vanadium flow batteries	2

第五章 钠电池技术（sodium battery technology）

章节序号 chapter number	章节名称 chapters	知识点 key points	课时 class hour
5.1	钠电池技术概述 overview of sodium battery technology	（1）钠电池的关键材料 （2）钠电池的经济和技术指标 （3）钠电池的未来发展展望 （1）key materials for sodium battery technology （2）economic and technical index of sodium battery technology （3）future development prospect of sodium battery technology	2
5.2	多种钠电池 classification of sodium battery technology	（1）钠硫电池 （2）zebra电池 （3）钠-空气电池 （4）钠离子电池 （1）sodium-sulfur batteries （2）zebra batteries （3）sodium-air batteries （4）sodium-ion batteries	2

第六章　镍氢电池技术（Ni-MH battery technology）

章节序号 chapter number	章节名称 chapters	知识点 key points	课时 class hour
6.1	镍氢电池介绍 introduction of Ni-MH batteries	（1）镍氢电池的储能原理 （2）镍氢电池的关键材料 （3）镍氢电池的功率和能量应用范围 （1）energy storage principle of Ni-MH batteries （2）key materials for Ni-MH batteries （3）power and energy application range of Ni-MH batteries	2
6.2	镍氢电池的发展 development of Ni-MH batteries	（1）镍氢电池的应用现状和产业链及环境问题 （2）镍氢电池相关新技术的发展 （3）镍氢电池的技术和经济指标及未来发展线路图 （1）application status, industrial chain, and environmental issues of Ni-MH batteries （2）development of new technology related to Ni-MH batteries （3）technical, economic indicators, and future development roadmap of Ni-MH batteries	2

第七章　超级电容器储能技术（supercapacitor energy storage technology）

章节序号 chapter number	章节名称 chapters	知识点 key points	课时 class hour
7.1	超级电容器概述 overview of supercapacitors	（1）超级电容器储能技术的基本原理 （2）超级电容器储能技术的发展历史 （1）basic principles of energy storage technology of supercapacitor （2）development history of energy storage technology of supercapacitor	2
7.2	超级电容器的关键材料与应用 key materials and application of supercapacitors	（1）多孔碳材料 （2）赝电容材料 （3）超级电容器电解液 （4）超级电容器的应用 （1）porous carbon materials （2）pseudocapacitive materials （3）electrolyte of supercapacitors （4）application of supercapacitors	2

第八章 储能技术在风力和光伏发电系统中的应用（application of energy storage technology in wind power and photovoltaic power generation system）

章节序号 chapter number	章节名称 chapters	知识点 key points	课时 class hour
8.1	风力发电和光伏发电技术概述及其对储能的需求 overview of wind and photovoltaic power generation technologies and their energy storage requirements	（1）风力发电和光伏发电技术概述 （2）风力发电和光伏发电技术对储能的需求 （1）overview of wind and photovoltaic power generation technologies （2）energy storage requirements of wind and photovoltaic power generation technologies	1
8.2	风力发电、光伏发电和储能技术的应用研究与未来发展 research on application and future development of wind power, photovoltaic power generation, and energy storage technologies	（1）风力发电和光伏发电系统中储能技术的应用研究 （2）风力发电、光伏发电和储能技术的未来发展 （1）research on application of energy storage technology in wind power generation and photovoltaic power generation system （2）future development of wind power, photovoltaic power generation, and energy storage technologies	1

第九章 储能技术在交通运输系统中的应用（application of energy storage technology in transportation system）

章节序号 chapter number	章节名称 chapters	知识点 key points	课时 class hour
9.1	储能技术在交通运输系统中的应用 application of energy storage technology in transportation system	（1）交通运输系统概述及其对储能技术的需求 （2）储能技术在交通运输系统中的应用现状 （1）overview of transportation systems and their requirements for energy storage technologies （2）application status of energy storage technology in transportation system	2

第十章 储能技术在电力系统中的应用（application of energy storage technology in power system）

章节序号 chapter number	章节名称 chapters	知识点 key points	课时 class hour
10.1	电力系统应用储能技术概述 overview of energy storage technology applied in power system	（1）电力系统应用储能技术的需求和背景 （2）储能技术在电力系统中的应用现状 （1）demand and background of energy storage technology applied in power system （2）application status of energy storage technology in power system	2

章节序号 chapter number	章节名称 chapters	知识点 key points	课时 class hour
10.2	电力系统应用储能技术的发展 development of energy storage technology applied in power system	（1）我国电力系统的储能应用实践 （2）适合电力系统应用的储能技术评价 （3）储能在电力系统应用中的发展趋势和重点研发方向 （1）application practice of energy storage in power system in China （2）evaluation of energy storage technologies suitable for power system application （3）development trend and key research directions of energy storage in power system application	2

第十一章　储能应用的经济性分析（economic analysis of energy storage applications）

章节序号 chapter number	章节名称 chapters	知识点 key points	课时 class hour
11.1	储能市场的现状及预期 current situation and perspectives of energy storage market	（1）储能市场的现状 （2）储能市场的预期 （1）current situation of energy storage market （2）perspectives of energy storage market	2
11.2	储能市场的应用 applications of energy storage market	（1）储能的应用 （2）储能电力服务叠加 （3）对储能电力应用服务的价值评估 （4）对储能应用的成本评估 （5）储能发展的主要瓶颈:成本 （6）储能成本降低的主要途径 （1）applications of energy storage （2）service stack of energy storage power （3）value assessment of energy storage power application services （4）cost assessment for energy storage applications （5）the main bottleneck of energy storage development: cost （6）the main way to reduce the cost of energy storage	2

4.10　"储能原理Ⅱ"教学大纲

课程名称：储能原理Ⅱ

Course Title：Energy Storage Principle Ⅱ

先修课程：高等数学Ⅰ，大学物理（含实验），工程力学，储能热流基础（含实验）

Prerequisites：Advanced Mathematics Ⅰ, Physics and Physics Experiments, Engineering Mechanics, Thermal–Fluid Fundamentals for Energy Storage（with Experiments）

学分：2

Credits：2

4.10.1　课程目的和基本内容（Course Objectives and Basic Contents）

本课程是储能科学与工程专业的一门专业基础必修课。

本课程的任务是使学生获得机械储能和热质储能方面的基本理论、基本知识和基本技能，培养学生分析问题和解决问题的能力，为以后深入学习储能领域中的内容以及从事相关科研工作打下基础。储能原理Ⅱ主要介绍机械及热质储能的基础理论、储能介质、储能过程性能分析及强化、储能系统设计及性能评价等，涉及的储能技术包括抽水蓄能、飞轮蓄能、显热储热、相变储热、热化学储热及压缩空气储能等。

储能原理Ⅱ开设的实验以机械储能和热质储能知识为背景。通过演示实验、科研平台参观、实验设备操作、实验结果分析及讨论等过程，使学生对储能系统的构成和工作原理有直观认识，培养学生的动手能力和分析问题、解决问题的综合能力。

This course is a mandatory fundamental course for undergraduates majoring in energy storage science and engineering.

This course enables students to acquire basic theories, basic knowledge, and basic skills in mechanical energy storage and heat and mass energy storage. It develops students' ability to analyze and solve problems, and lay a solid foundation for study and research in energy storage. Energy storage principle Ⅱ mainly introduces the basic theory, energy storage medium, performance analysis and enhancement of energy storage process, design and evaluation of energy storage system for mechanical energy storage and heat and mass energy storage, etc. The energy storage technologies involved in this course include pumped-hydro energy storage, flywheel energy storage, sensible heat storage, phase change heat storage, thermochemical heat storage and compressed air energy storage, etc.

The experiment of energy storage principle Ⅱ is based on the knowledge of mechanical energy storage and heat and mass energy storage. Through the demonstrative experiment, scientific research platform visit, experimental equipment operation, experimental results analysis and discussion, students can have an intuitive understanding of the composition, working principle of energy storage system, and cultivate their hands-on ability and comprehensive ability of analyzing and solving problems.

4.10.2　课程基本情况（Basic Information of the Course）

课程名称	储能原理II Energy Storage Principle II								
开课时间	一年级		二年级		三年级		四年级		总学分
	秋	春	秋	春	秋	春	秋	春	

课程名称	储能原理II Energy Storage Principle II			
开课时间	一年级／二年级／三年级／四年级 秋 春 秋 春 秋 春 秋 春	总学分	2	总学时 40
课程定位	储能科学与工程专业本科生储能基础课程群必修课			
授课学时分配	课堂讲授36学时+实验4学时			
先修课程	工科数学分析,大学物理(含实验),工程力学,储能热流基础(含实验)			
后续课程	可再生能源及其发电技术,储能系统设计,热质储能技术及应用(含实验),先进热力系统技术及仿真,流体机械原理及其储能应用,储能系统检测与估计			
教学方式	课堂教学,大作业,实验报告,平时作业			
考核方式	课程结束笔试成绩占60%,大作业及平时成绩占20%,实验占20%			
参考教材	[1] 丁玉龙,来小康,陈海生.储能技术及应用.北京:化学工业出版社,2018. [2] 黄志高.储能原理与技术.北京:中国水利水电出版社,2018.			
参考资料	[1] 张鸣远.流体力学.北京:高等教育出版社,2015. [2] 傅秦生,赵小明,唐桂华.热工基础与应用.3版.北京:机械工业出版社,2015. [3] 张仁元.相变材料与相变储能技术.北京:科学出版社,2009.			

4.10.3　教学目的和基本要求（Teaching Objectives and Basic Requirements）

（1）使学生明确机械储能和热质储能技术方面的基本概念、工作原理和应用现状，建立储能方面的基础知识体系。

（2）培养学生掌握不同储能技术的工作性能分析方法和性能强化手段，具体包括抽水蓄能、飞轮蓄能、显热储热、相变储热、热化学储热及压缩空气储能等。

（3）培养学生针对具体的储能问题，运用基础理论、技术和方法解决实际问题的能力，培养科学素养和系统思维能力，为将来从事相关科研工作铺垫坚实基础。

（1）Make students understand the basic concept, working principle and application status of mechanical energy storage and heat and mass energy storage technology, and establish the basic knowledge system of energy storage.

（2）Cultivate students to master the performance analysis and enhancement methods of different energy storage technologies, including pumped-hydro energy, flywheel energy storage, sensible heat storage, phase change heat storage, thermochemical heat storage and compressed air energy storage

（3）Cultivate students to obtain the ability of solving practical problems by using basic theories, technologies and methods according to specific energy storage problems, and have scientific literacy and systematic thinking ability, so as to lay a solid foundation for related scientific research work in the future.

4.10.4　课程大纲和知识点（Syllabus and Key Points）

第一章　绪论（introduction）

章节序号 chapter number	章节名称 chapters	知识点 key points	课时 class hour
1.1	储能背景 background of energy storage	（1）能源利用现状 （2）储能技术的必要性 （1）current situation of energy utilization （2）necessity of energy storage technology	0.5
1.2	储能技术的分类及其发展历程 classification and development of energy storage technology	（1）储能技术的分类 （2）储能技术的发展历程 （1）classification of energy storage technology （2）development of energy storage technology	1
1.3	机械储能的特点和应用现状 characteristics and application status of mechanical energy storage	（1）机械储能的特点 （2）机械储能的应用现状 （1）characteristics of mechanical energy storage （2）application status of mechanical energy storage	1
1.4	热质储能的特点和应用现状 characteristics and application status of heat and mass energy storage	（1）热质储能的特点 （2）热质储能的应用现状 （1）feature of heat and mass energy storage （2）application status of heat and mass energy storage	0.5

第二章　机械及热质储能理论基础（fundamentals of mechanical and heat and mass energy storage）

章节序号 chapter number	章节名称 chapters	知识点 key points	课时 class hour
2.1	机械储能的理论基础 fundamentals of mechanical energy storage	（1）机械储能相关基础理论：管路水力分析，伯努利方程 （2）机械储能相关基础理论：机电能量转换，动量矩定理，轴承技术 （1）mechanical energy storage related theories: hydraulic analysis of pipelines, Bernoulli's equation （2）mechanical energy storage related theories: mechanical and electrical energy conversion, theorem of moment of momentum, bearing technology	3

续表

章节序号 chapter number	章节名称 chapters	知识点 key points	课时 class hour
2.2	热质储能的热流原理 fundamentals of heat and mass energy storage	（1）热质储能相关基础理论：热能传递基本方式，流动相似原理，缩放喷管流动，边界层 （2）热质储能相关基础理论：物质相变理论，量热技术 （1）heat and mass energy storage related theories: basic modes of heat transfer, flow similarity criterion, laval nozzle, boundary layer （2）heat and mass energy storage related theories: theory of phase change, calorimetry	3

第三章 抽水蓄能技术（pumped-hydro energy storage technology）

章节序号 chapter number	章节名称 chapters	知识点 key points	课时 class hour
3.1	抽水蓄能的概念和原理 concept and theory of pumped-hydro energy storage	（1）抽水蓄能的概念 （2）抽水蓄能的原理 （1）concept of pumped-hydro energy storage （2）theory of pumped-hydro energy storage	1
3.2	抽水蓄能系统 system of pumped-hydro energy storage	（1）抽水蓄能系统简介 （2）抽水蓄能设备 （3）抽水蓄能系统设计 （1）introduction to pumped-hydro energy storage system （2）devices of pumped-hydro energy storage （3）design of pumped-hydro energy storage system	3

第四章 飞轮蓄能技术（flywheel energy storage technology）

章节序号 chapter number	章节名称 chapters	知识点 key points	课时 class hour
4.1	飞轮蓄能的概念和原理 concept and theory of flywheel energy storage	（1）飞轮蓄能的概念 （2）飞轮蓄能的原理 （1）concept of flywheel energy storage （2）theory of flywheel energy storage	1

续表

章节序号 chapter number	章节名称 chapters	知识点 key points	课时 class hour
4.2	飞轮蓄能系统 system of flywheel energy storage	（1）飞轮蓄能系统 （2）飞轮蓄能设备 （3）飞轮蓄能系统设计 （1）system of flywheel energy storage （2）devices of flywheel energy storage （3）design of flywheel energy storage system	3

第五章 显热储热技术（sensible heat storage technology）

章节序号 chapter number	章节名称 chapters	知识点 key points	课时 class hour
5.1	显热储热的概念和原理 concept and theory of sensible heat energy storage	（1）显热储热的概念 （2）显热储热的原理 （1）concept of sensible heat energy storage （2）theory of sensible heat energy storage	1
5.2	显热储热材料 sensible heat energy storage materials	（1）显热储能材料的分类 （2）显热储能材料的性能表征 （3）显热储能材料的性能强化 （1）classification of sensible heat storage materials （2）performance characterization of sensible heat storage materials （3）performance enhancement of sensible heat storage materials	1
5.3	显热储热过程的性能分析及强化 performance analysis and enhancement of sensible heat storage process	（1）显热储热过程的性能分析 （2）显热储热过程的性能强化 （1）performance analysis of sensible heat storage process （2）performance enhancement of sensible heat storage process	1
5.4	显热储热系统设计 design of sensible heat energy storage system	（1）显热储热设备 （2）显热储热系统设计 （1）devices of sensible heat energy storage （2）design of sensible heat energy storage system	1

第六章　相变储热技术（phase change heat storage technology）

章节序号 chapter number	章节名称 chapters	知识点 key points	课时 class hour
6.1	相变储热的概念和原理 concept and theory of phase change heat storage	（1）相变储热系统的概念 （2）相变储热系统的原理 （1）concept of phase change heat storage （2）theory of phase change heat storage	1
6.2	相变储热材料 phase change heat storage materials	（1）相变储热材料的分类 （2）相变储热材料的性能表征 （3）相变储热材料的性能强化 （1）classification of phase change heat storage materials （2）characterization of phase change heat storage materials （3）performance enhancement of phase change heat storage materials	1
6.3	相变储热过程的性能分析及强化 performance analysis and enhancement of phase change heat storage process	（1）相变储热过程的性能分析 （2）相变储热/放热过程的性能强化 （1）performance analysis of phase change heat storage process （2）performance enhancement of phase change heat storage/release processes	2
6.4	相变储热系统设计 design of phase change heat storage system	（1）相变储热设备 （2）相变储热系统设计 （1）devices of phase change heat storage （2）design of phase change heat storage system	1

第七章　热化学储热技术（thermochemical heat storage technology）

章节序号 chapter number	章节名称 chapters	知识点 key points	课时 class hour
7.1	热化学储热的概念和原理 concept and theory of thermochemical heat storage	（1）热化学储热系统的概念 （2）热化学储热系统的原理 （1）concept of thermochemical heat storage （2）theory of thermochemical heat storage	1
7.2	热化学储热材料 thermochemical heat storage materials	（1）热化学储能材料及其分类 （2）热化学储能材料的性能及其强化 （1）thermochemical energy storage materials and their classification （2）characterization and performance enhancement of thermochemical energy storage materials	1

续表

章节序号 chapter number	章节名称 chapters	知识点 key points	课时 class hour
7.3	热化学储热过程性能分析及强化 performance analysis and enhancement of thermochemical heat storage process	（1）热化学储热过程的性能分析 （2）热化学储热/放热过程的性能强化 （1）performance analysis of thermochemical heat storage process （2）performance enhancement of thermochemical heat storage/release processes	2
7.4	热化学储热系统设计 design of thermochemical heat storage system	（1）热化学储热系统的性能评价 （2）热化学储能系统设计 （1）performance evaluation of thermochemical heat storage system （2）design of thermochemical heat storage system	1

第八章 压缩空气储能技术（compressed air energy storage technology）

章节序号 chapter number	章节名称 chapters	知识点 key points	课时 class hour
8.1	压缩空气储能技术简介 introduction of compressed air energy storage technology	（1）压缩空气储能技术的概念 （2）压缩空气储能技术的工作原理 （1）concept of compressed air energy storage technology （2）operating principle of compressed air energy storage technology	1
8.2	压缩空气储能技术工作过程分析 working process analysis of compressed air energy storage technology	（1）压缩空气储能技术的工作过程分析 （2）工作过程能量转换 （1）working process analysis of compressed air energy storage technology （2）energy conversion in working process	2
8.3	压缩空气储能系统的性能评价 performance evaluation of compressed air energy storage technology	（1）压缩空气储能技术的性能评价指标 （2）压缩空气储能技术的系统设计 （1）performance evaluation index of compressed air energy storage technology （2）system design of compressed air energy storage technology	1
8.4	压缩空气储能技术的现状及应用前景 present situation and application prospect of compressed air energy storage technology	（1）压缩空气储能技术的现状 （2）压缩空气储能技术的应用前景 （1）present situation of compressed air energy storage technology （2）application prospect of compressed air energy storage technology	1

4.10.5　实验环节（Experiments）

序号 number	实验内容 experiment content	知识点 key points	课时 class hour
1	演示实验 experiment demonstration	（1）飞轮蓄能虚拟演示实验 （2）显热储能演示实验 （3）储能科研平台演示参观 （1）virtual demonstration experiment of flywheel energy storage （2）demonstrative experiment of sensible heat energy storage （3）energy storage research platform visit	1
2	抽水蓄能实验（二选一） pumped-hydro energy storage experiment（alternative）	（1）抽水蓄能虚拟演示实验 （2）抽水蓄能系统设计 （3）泵和管道等设备参数计算 （4）实验设备选配 （5）系统参数测试及数据分析 （6）抽水蓄能系统性能讨论 （1）virtual demonstrative experiment of pumped-hydro energy storage （2）system design of pumped-hydro energy storage （3）parameters calculation of pump and pipeline （4）selection of experimental device （5）system parameters test and data analysis （6）performance discussion of pumped-hydro energy storage system	3
3	相变储能实验（二选一） phase change energy storage experiment（alternative）	（1）相变储能虚拟演示实验 （2）相变储能系统设计 （3）工质和设备参数计算 （4）实验设备选配 （5）系统参数测试及数据分析 （6）相变储能系统性能讨论 （1）virtual demonstrative experiment of phase change energy storage （2）system design of phase change energy storage （3）parameter calculation of phase change energy storage material and devices （4）selection of experimental devices （5）system parameters test and data analysis （6）performance discussion of phase change energy storage system	3

热质储能课程群

5.1 "储能热流基础(含实验)"教学大纲

课程名称：储能热流基础（含实验）

Course Title：Fundamentals of Thermal-fluid Science in Energy Storage （with Experiments）

先修课程：高等数学Ⅰ，大学物理（含实验）

Prerequisites：Advanced Mathematics Ⅰ，Physics and Physics Experiments

学分：4.5

Credits：4.5

5.1.1 课程目的和基本内容（Course Objectives and Basic Contents）

本课程是储能科学与工程专业本科生的专业必修课程，是热质储能技术的理论和方法基础课程。主要目的及基本内容如下：

讲述工程热力学的基本概念，热力学第一定律和第二定律，气体、蒸气和湿空气的热力性质，气体的热力过程，热功转换设备和装置的热力分析及能量的合理利用等；讨论流体的主要物理性质，流体静力学、流体动力学的基本概念、基本原理、基本计算方法，理想流体动力学基础，理解相似理论与量纲分析的一般原理，管道阻力及管路计算，可压缩流动等。结合热质储能背景，通过对能量转化及流体流动规律的详细介绍，为学生进一步理解储能及用能过程中的质量传递及能量传递与转换过程奠定基础。

This course is a professional fundamental course for undergraduates majoring in engergy storage science and engineering, which covers the important theoretical and methodological foundation of heat and mass energy storage technology. The main purposes and basic contents include:

Introduces the basic concepts of engineering thermodynamics, the first and second laws of thermodynamics, the thermodynamic properties of gas, steam and wet air, the thermodynamic of the gas, thermodynamic of heat to power conversion equipments and devices and reasonable use of energy, etc. This course also discusses the main physical properties of fluids, the basic concepts, basic principles, and basic calculation methods of hydrostatics and fluid dynamics and foundation of ideal fluid movement; understanding the general principles of similarity theory and dimensional analysis, flow resistance and pipeline calculation, foundation of compressible flow, etc. Combined with the background of heat and mass energy storage, through a detailed introduction to the law of energy conversion and fluid flow, this course will lay the foundation for students to further understand the various thermodynamic systems, and mass transfer and energy transfer and conversion processes in the process of energy storage and usage.

5.1.2　课程基本情况（Basic Information of the Course）

课程名称	储能热流基础（含实验） Fundamentals of Thermal-fluid Science in Energy Storage（with Experiments）										
开课时间	一年级		二年级		三年级		四年级		总学分	4.5	总学时	80
	秋	春	秋	春	秋	春	秋	春				
课程定位	储能科学与工程专业本科生热质储能课程群必修课											
授课学时分配	课堂讲授64学时+实验16学时											
先修课程	高等数学Ⅰ,大学物理（含实验）											
后续课程	储能原理II,传热传质学,先进热力系统技术及仿真,可再生能源及其发电技术,流体机械原理及其储能应用,氢能储存与应用,热质储能综合实验											
教学方式	课堂教学,大作业,平时作业											
考核方式	课程结束笔试成绩占70%,平时成绩占30%											
参考教材	[1] 朱明善,刘颖,林兆庄,等. 工程热力学. 2版. 北京:清华大学出版社,2011. [2] 张鸣远. 流体力学. 北京:高等教育出版社,2010.											
参考资料	[1] 何雅玲.工程热力学精要解析. 西安:西安交通大学出版社,2014. [2] Cengel Y A, Boles M A. Thermodynamics:An Engineering Approach. 6th ed. 何雅玲,缩编. 北京:电子工业出版社,2009.											

5.1.3　教学目的和基本要求（Teaching Objectives and Basic Requirements）

（1）掌握与热质储能密切相关的工程热力学、流体力学基本概念和术语，能够准确理解其内涵和工程意义。

（2）掌握热力学第一和第二定律，能够对热力过程和循环进行能量分析。

（3）掌握工质热力性质的获取及计算，能够利用公式、手册或软件得到工程上常用工质的热力性质。

（4）掌握流体的主要物理性质和力学性质；掌握流体平衡状态下的压强分布规律及压强计算。

（5）掌握流体流动遵循的基本规律。

（6）掌握量纲分析方法、相似理论和相似准则。

（1）Master the basic concepts and terminology of engineering thermodynamics and fluid mechanics closely related to heat and mass energy storage, and accurately understand its connotation and engineering significance.

（2）Master the first and second laws of thermodynamics, and be able to perform energy analysis on thermodynamic processes and cycles.

（3）Master the calculation and acquisition of thermodynamic properties of working fluid, and be able to use formulas, manuals or softwares to obtain the properties of working fluids that are commonly used in engineering.

（4）Master the main physical properties and various mechanical properties of fluids. Master the pressure distribution law and pressure calculation under equilibrium state.

（5）Master the basic laws of fluid flow.

（6）Master dimensional analysis method, similarity theory and similarity criterion.

5.1.4 课程大纲和知识点（Syllabus and Key Points）

第一部分　工程热力学

第一章　绪论（introduction）

章节序号 chapter number	章节名称 chapters	知识点 key points	课时 class hour
1.1	热力学与流体力学的内涵 connotation of thermodynamics and fluid mechanics	（1）热力学与环境 （2）热力学与生活 （3）流体之美 （4）热力学及流体力学的研究方法 （1）thermodynamics and environment （2）thermodynamics and life （3）beauty of fluid （4）research methods of thermodynamics and fluid mechanics	2
1.2	热力学及流体力学发展简史 brief history of the development of thermodynamics and fluid mechanics	（1）热力学简史 （2）流体力学简史 （1）a brief history of thermodynamics （2）a brief history of fluid mechanics	

续表

章节序号 chapter number	章节名称 chapters	知识点 key points	课时 class hour
1.3	储能过程中的热力学及流体力学 thermodynamics and fluid mechanics in energy storage process	（1）热力学在储能过程中的应用 （2）流体力学在储能过程中的应用 （1）application of thermodynamics in energy storage process （2）application of fluid mechanics in the process of energy storage	2

第二章 热力学的基本概念（basic concepts of thermodynamics）

章节序号 chapter number	章节名称 chapters	知识点 key points	课时 class hour
2.1	热力系统 thermodynamic system	（1）系统及分类 （2）工质 （1）system and its classification （2）working fluid	2
2.2	状态及状态参数 state and state parameters	（1）状态 （2）状态参数 （3）基本状态参数 （1）state （2）state parameters （3）basic state parameters	
2.3	平衡状态及状态方程式 equilibrium state and state equation	（1）平衡状态 （2）状态方程式 （3）状态参数坐标图 （1）equilibrium state （2）state equation （3）state parameter coordinate graph	2
2.4	热力过程 thermodynamic process	（1）准静态过程 （2）可逆过程 （1）quasi-equilibrium process （2）reversible process	
2.5	热力循环 thermodynamic cycle	（1）动力循环 （2）制冷循环/热泵循环 （1）power cycle （2）refrigeration cycle / heat pump cycle	

第三章 热力学第一定律（the first law of thermodynamics）

章节序号 chapter number	章节名称 chapters	知识点 key points	课时 class hour
3.1	热力学第一定律的实质 essence of the first law of thermodynamics	（1）热力学第一定律的描述 （2）能量守恒 （1）description of the first law of thermodynamics （2）energy conservation	2
3.2	热力学能和总能 internal energy and total energy	（1）热力学能 （2）总能 （1）internal energy （2）total energy	
3.3	能量的传递与转化，焓 energy transfer and conversion, enthalpy	（1）作功 （2）传热 （3）焓 （1）work （2）heat transfer （3）enthalpy	
3.4	热力学第一定律的基本表达式 basic expression of the first law of thermodynamics	（1）热力学第一定律的能量方程式 （2）闭口系能量方程式 （1）energy equation of the first law of thermodynamics （2）energy equation of close system	
3.5	开口系统能量方程 energy equation of open system	（1）开口系能量方程 （2）稳定流动能量方程 （1）energy equation of open system （2）energy equation of steady flow	2
3.6	能量方程的实际应用 application of energy equation	（1）动力机械 （2）换热器 （3）绝热节流 （1）power machine （2）heat exchanger （3）adiabatic throttling	

第四章 气体和蒸气的性质（properties of gas and vapor）

章节序号 chapter number	章节名称 chapters	知识点 key points	课时 class hour
4.1	理想气体的概念 concept of ideal gas	（1）理想气体 （2）理想气体状态方程式 （1）ideal gas （2）ideal gas equation of state	3

续表

章节序号 chapter number	章节名称 chapters	知识点 key points	课时 class hour
4.2	理想气体的比热容 specific heat capacity of ideal gas	（1）比热容 （2）迈耶公式 （3）真实比热容与平均比热容 （1）specific heat capacity （2）Mayer formula （3）true specific heat capacity and average specific heat capacity	3
4.3	理想气体的热力学能、焓及熵 internal energy, enthalpy and entropy of ideal gas	（1）热力学能 （2）焓 （3）熵 （1）internal energy （2）enthalpy （3）entropy	
4.4	水蒸气的饱和状态和相图 saturation state and phase diagram of vapor	（1）凝结和汽化 （2）相图 （1）condensation and vaporization （2）phase diagram	
4.5	水的定压产生过程 water vaporization process	（1）水的定压产生过程 （2）水和水蒸气的状态参数 （1）water vaporization process （2）state parameters of water and steam	3
4.6	水蒸气的表和图 figure and table of water vapor for gas and steam properties	（1）水蒸气表 （2）T–s图 （3）h–s图 （1）water vapor table （2）T–s diagram （3）h–s diagram	

第五章　气体的热力过程（thermodynamic process of gas）

章节序号 chapter number	章节名称 chapters	知识点 key points	课时 class hour
5.1	研究热力过程的目的及一般方法 purpose and general method of thermal process	气体的基本热力过程 basic thermal process of gas	0.5

续表

章节序号 chapter number	章节名称 chapters	知识点 key points	课时 class hour
5.2	理想气体的基本热力过程 basic process of ideal gas	（1）定压过程 （2）定容过程 （3）定温过程 （4）绝热过程 （1）isobaric process （2）isochoric process （3）isothermal process （4）adiabatic process	1.5
5.3	理想气体的多变过程 polytropic process of ideal gas	（1）理想气体的多变过程方程式 （2）多变过程的p-v图及T-s图 （3）多变过程功、技术功及过程热量 （1）polytropic process equation of ideal gas （2）p-v diagram and T-s diagram of polytropic process （3）work, technical work and heat of polytropic process	2
5.4	水蒸气的基本过程 basic process of water vapor	（1）水蒸气的定压过程 （2）水蒸气的定容过程 （3）水蒸气的定温过程 （4）水蒸气的绝热过程 （1）isobaric process of steam （2）isochoric process of steam （3）isothermal process of steam （4）adiabatic process of steam	

第六章　热力学第二定律（the second law of thermodynamics）

章节序号 chapter number	章节名称 chapters	知识点 key points	课时 class hour
6.1	热力学第二定律 the second law of thermodynamics	（1）热力过程的方向性 （2）热力学第二定律的表述 （1）directionality of thermodynamic processes （2）expression of the second law of thermodynamics	
6.2	卡诺循环及卡诺定理 Carnot cycle and Carnot principles	（1）卡诺循环 （2）概括性卡诺循环 （3）多热源的卡诺循环 （4）卡诺定理 （1）Carnot cycle （2）generalized Carnot cycle （3）Carnot cycle with multiple heat sources （4）Carnot principles	1.5

续表

章节序号 chapter number	章节名称 chapters	知识点 key points	课时 class hour
6.3	熵、热力学第二定律数学表达式 entropy, mathematical expressions of the second law of thermodynamics	（1）状态参数熵的导出 （2）热力学第二定律的数学表达 （3）孤立系统熵增原理 （1）derivation of state parameter entropy （2）mathematical expressions of the second law of thermodynamics （3）entropy increase principle of isolated system	1.5
6.4	热力学第二定律的应用 application of the second law of thermodynamics	应用 application	2

第七章 压气机的热力过程（thermodynamic process of the compressor）

章节序号 chapter number	章节名称 chapters	知识点 key points	课时 class hour
7.1	单级活塞式压气机的工作原理和耗功 working principle and power consumption of single stage compressor	（1）工作原理 （2）压气机理论功耗 （1）working principle （2）theoretical power consumption of compressor	1.5
7.2	余隙容积的影响 influence of clearance volume	（1）余隙容积 （2）生产量 （3）理论功耗 （1）clearance volume （2）production volume （3）theoretical power consumption	
7.3	多级压缩和级间冷却 multistage compression and intercooling	多级压缩、级间冷却压气机的基本原理及计算 basic principle and calculation of multi-stages compression and intercooling compressor	1.5
7.4	叶轮式压气机的工作原理 working principle of impeller compressor	叶轮式压气机的工作原理及计算 working principle and calculation of impeller compressor	

第八章 热力学在储能工程中的应用案例介绍（applications of theymodynamics in energy storage）（4 学时）

第二部分　流体力学

第一章　流体及其主要物理性质（fluid and its main physical properties）

章节序号 chapter number	章节名称 chapters	知识点 key points	课时 class hour
1.1	流体的物理性质 physical properties of fluid	（1）流体的黏性 （2）扩散 （3）连续介质假说 （4）流体的热力学性质 （5）作用在流体上的力 （6）表面张力 （1）viscosity of the fluid （2）diffusion （3）continuum hypothesis （4）thermodynamic properties of fluid （5）force acting on the fluid （6）surface tension	2
1.2	量纲与单位 base unit and units	（1）量纲一致性原理 （2）基本量纲与导出量纲 （1）dimensional consistency principle （2）basic dimension and derived dimension	

第二章　流体静力学（hydrostatics）

章节序号 chapter number	章节名称 chapters	知识点 key points	课时 class hour
2.1	流体静压强及其特性 hydrostatic pressure and its characteristics	（1）静压强 （2）理想流体 （1）hydrostatic pressure （2）ideal fluid	2
2.2	静止流体的平衡微分方程 equilibrium differential equation of stationary fluid	（1）静止流体的平衡微分方程推导 （2）流体的平衡条件 （3）等压面 （1）derivation of equilibrium differential equations of stationary fluids （2）balance condition of fluid （3）isobaric surface	2
2.3	重力场中静止流体的压强分布 pressure distribution of static fluid in gravity field	（1）液体中的压强分布 （2）气体中的压强分布 （1）pressure distribution in liquid （2）pressure distribution in gas	2

续表

章节序号 chapter number	章节名称 chapters	知识点 key points	课时 class hour
2.4	压强测量 pressure measurement	（1）绝对压强 （2）计示压强 （3）真空压强 （4）液柱式测压计 （1）absolute pressure （2）gauge pressure （3）vacuum pressure （4）manometer	2
2.5	相对静止流体内的压强分布 pressure distribution in relatively static fluid	等角速度旋转运动 constant angular velocity rotation	
2.6	作用在平面上的流体静压力 hydrostatic pressure acting on the plane	（1）作用在平面上的流体静压力 （2）不可压缩流体的平面运动 （1）hydrostatic pressure acting on the plane （2）planar motion of incompressible fluid	

第三章　流体运动概述（overview of fluid motion）

章节序号 chapter number	章节名称 chapters	知识点 key points	课时 class hour
3.1	描述流体运行的两种方法 two ways to describe fluid motion	（1）拉格朗日方法 （2）欧拉方法 （3）定常与非定常流动 （4）一维、二维和三维流动 （1）Lagrangian method （2）Euler method （3）steady and unsteady flows （4）one-dimensional, two-dimensional and three-dimensional flow	2
3.2	迹线、流线和脉线 pathline, streamline and streakline	（1）迹线、流线和脉线 （2）流管 （1）pathline, streamline and streakline （2）steam tube	2

续表

章节序号 chapter number	章节名称 chapters	知识点 key points	课时 class hour
3.3	物质导数 material derivative	（1）流体质点的加速度 （2）物质导数 （3）圆柱坐标系 （1）acceleration of fluid particle （2）material derivative （3）cylindrical coordinate system	2
3.4	流体微团运动分析 motion analysis of fluid particle	（1）线变形 （2）相对体积膨胀率 （3）流体微团的旋转 （4）剪切变形率 （1）line deformation （2）relative volume expansion rate （3）rotation of fluid particle （4）shear deformation rate	2
3.5	连续方程 continuity equation	（1）连续方程推导 （2）不可压缩流动与等密度流动 （3）一维流动的连续方程 （1）derivation of continuous equation （2）incompressible flow and equal density flow （3）continuity equation of one-dimensional flow	2

第四章 理想流体运动基础（foundation of ideal fluid movement）

章节序号 chapter number	章节名称 chapters	知识点 key points	课时 class hour
4.1	欧拉方程 Euler equation	（1）欧拉方程各项的含义 （2）直角坐标系下的欧拉方程 （1）meaning of the Euler equation （2）Euler equation in Cartesian coordinate system	2
4.2	自然坐标系中的欧拉方程 Euler equation in natural coordinate system	（1）流线坐标系 （2）欧拉方程在自然坐标系中的表达式 （1）streamline coordinate system （2）expression of Euler equation in the natural coordinate system	

续表

章节序号 chapter number	章节名称 chapters	知识点 key points	课时 class hour
4.3	伯努利方程 Bernoulli's equation	（1）自然坐标系下伯努利方程的推导 （2）伯努利方程的物理意义 （3）静压强、动压强和滞止压强 （4）总能头线与测压管线 （1）derivation of Bernoulli's equation in natural coordinate system （2）the physical meaning of Bernoulli's equation （3）static pressure, dynamic pressure and stagnation pressure （4）total energy head line and pressure measuring pipeline	2
4.4	伯努利方程在储能系统中的应用 application of Bernoulli's equation in energy storage system	（1）伯努利方程在抽水蓄能中的应用 （2）伯努利方程在流体测量中的应用 （1）application of Bernoulli's equation in pumped-hydro energy storage system （2）application of Bernoulli's equation in fluid measurement	2

第五章 量纲分析与动力相似（dimensional analysis and similarity of dynamics）

章节序号 chapter number	章节名称 chapters	知识点 key points	课时 class hour
5.1	量纲分析 dimensional analysis	量纲分析 dimensional analysis	
5.2	白金汉 π 定理 Buckingham's Pi theorem	（1）白金汉 π 定理及其应用 （2）动力相似 （3）模型实验 （1）Buckingham's Pi theorem and its application （2）dynamic similarity （3）model experiment	2

第六章 流动阻力及管路计算（flow resistance and pipeline calculation）

章节序号 chapter number	章节名称 chapters	知识点 key points	课时 class hour
6.1	流动阻力及管路计算 flow resistance and pipeline calculation	（1）摩擦阻力 （2）局部阻力 （3）流动的总阻力 （4）管路计算量纲分析 （1）friction resistance （2）local resistance （3）total resistance of flow （4）dimensional analysis of pipeline calculation	2

第七章 可压缩流动基础（foundation of compressible flow）

章节序号 chapter number	章节名称 chapters	知识点 key points	课时 class hour
7.1	可压缩流动的基本概念 basic concepts of compressible flow	（1）定常流动 （2）非定常流动 （3）声速 （4）马赫数 （5）亚声速流动、跨声速流动、超声速流动、极超声速流动 （6）等熵流动 （1）constant flow （2）unsteady flow （3）speed of sound （4）Mach number （5）subsonic flow, transonic flow, supersonic flow and hypersonic flow （6）isentropic flow	2
7.2	一维定常可压缩流动的基本方程 basic equation of one-dimensional steady compressible flow	（1）能量方程 （2）动量方程 （3）状态方程 （4）等熵过程方程式 （5）等熵滞止状态 （1）energy equation （2）momentum equation （3）state equation （4）isentropic process equation （5）isentropic stagnation state	2

续表

章节序号 chapter number	章节名称 chapters	知识点 key points	课时 class hour
7.3	激波关系式 shock wave relations	（1）正激波 （2）激波的形成与厚度 （1）normal shock （2）formation and thickness of shock wave	2
7.4	几何喷管中的流动 fluid flow in a geometric nozzle	（1）收缩形喷管 （2）缩放形喷管 （1）retractable nozzle （2）laval nozzle	

5.1.5 实验环节（Experiments）

序号 number	实验内容 experiment content	知识点 key points	课时 class hour
1	测量的基本知识 basic knowledge of measurement	（1）测量的误差 （2）随机误差分析 （3）系统的误差分析 （4）误差的合成、间接测量的误差传递与分配 （1）measurement error （2）random error analysis （3）analysis of system error （4）error synthesis, error propagation and distribution of indirect measurement	2
2	温度及热流量测量 temperature and heat flux measurement	（1）温度测量的基本概念 （2）膨胀式温度计 （3）热电偶温度计 （4）热电阻温度计 （5）辐射方法测温 （6）热流量测量 （1）basic concept of temperature measurement （2）expansion thermometer （3）thermocouple thermometer （4）resistance thermometer （5）temperature measurement by radiation method （6）heat flux measurement	2

续表

序号 number	实验内容 experiment content	知识点 key points	课时 class hour
3	湿度测量 humidity measurement	（1）空气湿度的表示方法 （2）干湿球法湿度测量 （3）露点法湿度测量 （4）吸湿法湿度测量 （5）饱和盐溶液湿度校正装置 （1）expression of air humidity （2）humidity measurement by dry and wet bulb method （3）humidity measurement by dew point method （4）humidity measurement by hygroscopic method （5）humidity correction device for saturated salt solution	2
4	压力测量 pressure measurement	（1）液柱式压力计 （2）弹性式压力计 （3）压力(差压)传感器 （4）压力检测仪表的选择与校验 （1）liquid column manometer （2）elastic pressure gauge （3）pressure (differential pressure) sensor （4）selection and calibration of pressure measuring instruments	2
5	流速及流量测量 velocity and flowrate measurement	（1）流量测量方法 （2）速度式流量测量方法 （3）容积式流量测量方法 （4）质量式流量测量方法 （5）差压式流量测量方法 （1）flowrate measurement method （2）velocity flowrate measurement method （3）volumetric flowrate measurement method （4）mass flowrate measurement method （5）differential pressure flowrate measurement method	2
6	流体静力学实验 hydrostatic experiment	（1）流体静压强测量 （2）流体静力学基本方程验证 （3）流体静力学现象实验分析 （1）hydrostatic pressure measurement （2）verification of basic hydrostatic equations （3）experimental analysis of hydrostatic phenomena	2

续表

序号 number	实验内容 experiment content	知识点 key points	课时 class hour
7	伯努利方程实验 experiment of Bernoulli's equation	（1）验证流体恒定流动时的伯努利方程 （2）流体能量转换特性 （3）流速、流量、压强等参数的实际测量 （1）verification of the Bernoulli equation for flowing fluid （2）fluid energy conversion characteristics （3）actual measurement of parameters such as flow velocity, flow, pressure, etc.	2
8	压气机实验 compressor experiment	（1）观察活塞式压气机的外形、部件 （2）回顾压气机的工作过程 （3）对活塞式压气机热力过程进行实验测试，记录状态点参数并对其性能进行计算 （1）observe the shape and components of the piston compressor （2）review the working process of the compressor （3）conduct experimental tests on the thermal process of the piston compressor, record the state point parameters and calculate its performance	2

5.2 "传热传质学"教学大纲

课程名称：传热传质学

Course Title：Heat and Mass Transfer

先修课程：高等数学Ⅰ，大学物理（含实验），储能热流基础（含实验）

Prerequisites：Advanced MathematicsⅠ，Physics and Physics Experiments，Fundamentals of Thermal-Fluid Science in Energy Storage（with Experiments）

学分：3

Credits：3

5.2.1 课程目的和基本内容（Course Objectives and Basic Contents）

本课程是一门专业基础课，是热、质储能技术的理论和方法基础。它综合运用理论分析、实验研究和数值模拟的方法，对各类传热、传质问题分析其物理过程的机理，建立相应的物理与数学模型，介绍主要的求解温度分布、浓度分布及所传递的热量与质量的计算方法，及通过实验获得的传热、传质实验关联式，以使学生获得比较宽广和坚实的热量、质量传递规律的基础知识和分析与计算储能工程中的各类传热、传质问题的能力。

传热传质学课程的主要学习内容包括：稳态与非稳态导热的理论及分析和数值求解方法；对流传热传质的研究方法，无相变对流传热及气液相变和固液相变对流传热的计算方法；辐射传热的基本知识及太阳能利用中的辐射传热问题；换热器的种类及热质设计方法；质扩散、对流传质及吸附过程基本规律和计算方法，及强化传热、传质过程的主要原理与技术。

　　This course is a professional fundamental course, which covers the important theoretical and methodological foundation of heat and mass energy storage technology. It comprehensively uses analytical, experimental, and numerical tools to study a variety of heat and mass transfer problems. The course content includes analyzing physical mechanisms of their heat and mass transfer process, establishing their physical and mathematical models, introducing solution methods for temperature distributions,concentration distributions, heat transfered and mass transfered and providing experimental heat and mass transfer correlations. By learning this course, the students will acquire a broad and solid knowledge of the fundamental principles of heat and mass transfer, and grasp the basic ability to analyze and calculate heat and mass transfer problems in energy storage engineering.

　　The main learning contents of heat and mass transfer course include: theory and analytical and numerical methods for steady and unsteady heat conduction; research methods of convective heat and mass transfer, methods for solving heat and mass transfer rate of single-phase convective heat transfer problem, liquid-vapor phase change heat transfer, and solid-liquid phase change heat transfer; basic knowledge of thermal radiative heat transfer, and its application in solar energy utilization; types of heat-mass exchangers and their thermo-mass design methods; basic laws of mass diffusion, mass convection, mass adsorption, and corresponding calculation methods; principles and technologies of enhancing heat and mass transfer processes.

5.2.2　课程基本情况（Basic Information of the Course）

课程名称	传热传质学 Heat and Mass Transfer								
开课时间	一年级		二年级		三年级		四年级		总学分	3	总学时	56
	秋	春	秋	春	秋	春	秋	春				
课程定位	储能科学与工程专业本科生热质储能课程群必修课								
授课学时分配	课堂讲授56学时+上机学时8学时（上机学时不含在总学时中）								
先修课程	高等数学Ⅰ,大学物理（含实验）,储能热流基础（含实验）								
后续课程	热质储能技术及应用（含实验）,热质储能综合实验								
教学方式	课堂教学,大作业,讨论,平时作业								
考核方式	课程结束笔试成绩占70%,上机报告占10%,平时成绩占20%								

参考教材	陶文铨. 传热学. 5版. 北京: 高等教育出版社, 2019.
参考资料	[1] 丁玉龙, 来小康, 陈海生. 储能技术及应用. 北京: 化学工业出版社, 2018. [2] 王秋旺. 传热学重点难点及典型题精解. 西安: 西安交通大学出版社, 2001. [3] Cengel Y A, Ghajar A J. Heat and Mass Transfer: Fundamentals and Applications. 5th ed. New York: McGraw-Hill Companies, Inc., 2015. [4] Holman J P. Heat Transfer. 10th ed. New York: McGraw-Hill Companies, Inc., 2010. [5] Bergman T L, Lavine A S, Incropera F P, et al. Introduction to Heat Transfer. 6th ed. Hoboken: John Wily & Sons, Inc., 2011.

5.2.3　教学目的和基本要求（Teaching Objectives and Basic Requirements）

（1）使学生获得比较宽广和坚实的热量、质量传递规律的基础知识;

（2）使学生具备分析储能工程中的传热、传质问题的基本能力;

（3）使学生掌握计算储能工程中的传热、传质问题的基本方法;

（4）使学生具备对热质储能中扩散型传热、传质问题建立物理模型和数值计算的能力。

（1）The students will acquire broad and solid knowledge of the fundamental principles of heat and mass transfer.

（2）The students will have basic ability to analyze heat and mass transfer problems in energy storage engineering.

（3）The students will grasp the basic methods for solving heat and mass transfer problems in energy storage engineering.

（4）The students will be able to establish physical models and conduct numerical simulation for diffusion-type heat and mass transfer problems in energy storage engineering.

5.2.4　课程大纲和知识点（Syllabus and Key Points）

第一章　绪论（introduction）

章节序号 chapter number	章节名称 chapters	知识点 key points	课时 class hour
1.1	传热传质学及其研究内容 heat and mass transfer and their research contents	（1）传热传质学的基本定义 （2）传热传质学应用领域 （1）what is heat and mass transfer （2）application areas of heat and mass transfer	0.5

续表

章节序号 chapter number	章节名称 chapters	知识点 key points	课时 class hour
1.2	热量传递的三种方式 three modes of heat transfer	（1）导热 （2）对流 （3）辐射 （1）conduction （2）convection （3）radiation	1
1.3	传热过程与传热系数 heat transfer process and heat transfer coefficient	（1）传热过程的概念 （2）传热过程的计算 （3）热阻概念及分析 （1）concept of overall heat transfer process （2）calculations of overall heat transfer process （3）concept and analysis of thermal resistance	1
1.4	传热学研究方法及传热学在储能中的应用 methods of research for heat transfer and applications of heat transfer in energy storage engineering	（1）传热学研究方法 （2）传热学在储能工程中的应用 （1）methods of research for heat transfer （2）applications of heat transfer in energy storage engineering	0.5

第二章 稳态热传导的规律及计算（fundamental laws and solution of steady state heat conduction problems）

章节序号 chapter number	章节名称 chapters	知识点 key points	课时 class hour
2.1	导热基本定律：傅里叶定律 fundamental laws of heat conduction：Fourier's law	（1）温度场 （2）等温线（面） （3）温度梯度 （4）傅里叶定律 （5）导热系数 （1）temperature field （2）isotherm (surface) （3）temperature gradient （4）Fourier's law （5）thermal conductivity	1
2.2	导热问题的数学描写 mathematical formulation of heat conduction problem	（1）导热微分方程式 （2）定解条件 （1）differential equation for heat conduction （2）conditions for unique solutions	0.5

续表

章节序号 chapter number	章节名称 chapters	知识点 key points	课时 class hour
2.3	典型一维稳态导热问题的分析解 analytical solution of typical one-dimensional steady-state heat conduction problem	（1）通过平壁的稳态导热 （2）通过圆筒壁的导热 （3）通过球壳的导热 （1）steady-state heat conduction in plane wall （2）heat conduction in cylinder wall （3）heat conduction in spherical shell	0.5
2.4	通过肋片的导热 heat conduction in fins	（1）等截面肋片的导热 （2）肋效率 （3）肋片换热量的计算 （1）heat conduction in fins with uniform cross-section （2）fin efficiency （3）calculation of heat transfer in fins	1
2.5	具有内热源的一维导热问题 one-dimensional steady-state heat conduction with internal heat source	（1）具有内热源的平板导热 （2）具有内热源的圆柱导热 （1）heat conduction in plates with internal heat source （2）heat conduction in cylinders with internal heat source	0.5
2.6	多维稳态导热的求解 solution of multidimensional steady-state heat conduction problems	（1）分离变量法 （2）形状因子 （1）method of separation of variables （2）shape factor	0.5
2.7	储能材料等效导热系数的确定 determination of effective thermal conductivity of the energy storage material	（1）等效导热系数 （2）等效模型 （1）equivalent thermal conductivity （2）equivalent model	1

第三章　非稳态热传导的计算（calculation of unsteady heat conduction）

章节序号 chapter number	章节名称 chapters	知识点 key points	课时 class hour
3.1	非稳态导热的基本概念 basic concept of unsteady heat conduction	（1）非稳态导热的基本特点 （2）热扩散率 （3）内、外热阻之比对温度分布的影响 （1）basic characteristics of unsteady heat conduction （2）thermal diffusivity （3）effect of ratio of internal and external thermal resistances on temperature distribution	0.5

续表

章节序号 chapter number	章节名称 chapters	知识点 key points	课时 class hour
3.2	零维问题的分析法:集总参数法 analysis of zero- dimension problems: lumped parameter method	（1）集总参数法的方法实质和方法要点 （2）集总参数法的适用条件 （3）热电偶时间常数 （1）essence and main points of lumped parameter method （2）applicable conditions of lumped parameter method （3）time constant of thermocouple	1
3.3	典型一维物体非稳态导热的分析解 analytical solution of typical one-dimensional unsteady heat conduction problem	（1）一维非稳态导热物理问题的数学描写 （2）正规状况阶段非稳态导热的计算方法 （1）mathematical formulation of the physical problem of one-dimensional unsteady heat conduction （2）calculation method of unsteady heat conduction in regular regime	1
3.4	半无限大物体的非稳态导热 unsteady heat conduction in a semi-infinite object	（1）半无限大的物理概念 （2）吸热系数 （3）半无限大物体温度场的解析解 （1）physical conception of semi-infinite body （2）heat absorption coefficient （3）analytical solution of semi-infinite model temperature field	0.5
3.5	简单几何形状多维非稳态导热的分析解 analytical solution of multi-dimensional unsteady heat conduction in simple geometric objects	（1）多维非稳态导热乘积解的理论依据及解的形式 （2）乘积解的适用条件 （1）theoretical basis and solution form of the product solution for the multidimensional unsteady heat conduction problems （2）applicable conditions of the product solution method	0.5
3.6	两相一维非稳态导热过程及其在储能中的应用 one-dimensional two-phase unsteady thermal conduction process and its application in energy storage	（1）两相一维非稳态导热微分方程 （2）相界面处的连续条件 （1）one-dimensional differential equation for two-phase unsteady heat conduction （2）continuity conditions at the interface boundary	1
3.7	非傅里叶导热问题简介 a brief introduction to non-Fourier heat conduction problems	（1）非傅里叶导热的四种情形 （2）松弛时间 （1）four cases of non-Fourier conduction （2）relaxation time	0.5

第四章　热传导问题的数值解法（numerical solution of heat conduction problem）

章节序号 chapter number	章节名称 chapters	知识点 key points	课时 class hour
4.1	导热问题数值求解基本思想 fundamental concept of numerical solution of heat conduction problem	（1）数值解法的本质 （2）数值解法的基本步骤 （1）essence of numerical method （2）basic steps of numerical method	0.5
4.2	内节点离散方程的建立 establishment of discrete equations for internal nodes	（1）离散的概念 （2）泰勒展开法 （3）热平衡法 （1）concept of discretization （2）Tayler series expansion （3）heat balance method	1
4.3	边界节点离散方程的建立及代数方程的求解 establishment of discrete equations for boundary nodes and solution of algebraic equation	（1）内节点与边界节点 （2）对流边界与绝热边界的处理 （3）求解代数方程的迭代法 （4）迭代收敛的概念 （5）肋片温度场计算 （1）inner node and boundary node （2）numerical treatment of convective and insolated boundary nodes （3）iterative method for solving algebraic equations （4）concept of iterative convergence （5）numerical solution of fin temperature field	1.5
4.4	非稳态导热的数值解法 numerical solution of unsteady heat conduction problems	（1）非稳态导热的概念 （2）显式与隐式 （3）斯特藩问题的数值计算 （1）concept of unsteady heat conduction （2）explicit and implicit schemes （3）numerical solution of Stefan problem	1.5
4.5	数值计算的稳定性、收敛性及精度 stability, convergence and accuracy of numerical solution	（1）数值计算的稳定性 （2）数值计算的收敛性 （3）数值解的精度 （1）stability of numerical solution （2）convergence of numerical solution （3）accuracy of numerical solution	0.5

第五章 对流传热的理论分析及实验研究基础（fundamentals of theoretical analysis and experimental study of convective heat transfer）

章节序号 chapter number	章节名称 chapters	知识点 key points	课时 class hour
5.1	对流换热概述 overview of convective heat transfer	（1）对流换热的概念 （2）影响对流换热的因素 （3）对流换热的分析及研究方法 （4）表面传热系数的计算 （1）concept of convective heat transfer （2）factors affecting convective heat transfer （3）analysis and research methods of convective heat transfer （4）calculation of convective heat transfer coefficient	1
5.2	对流换热数学描写 mathematical formulation of convective heat transfer	（1）导出对流换热微分方程组的理论依据 （2）能量微分方程的导出 （1）theoretical basis for the derivation of differential equations of convective heat transfer （2）derivation of energy differential equation	1
5.3	边界层对流换热的数学描写 mathematical formulation of convective heat transfer in boundary layer	（1）边界层的概念 （2）边界层的主要特点及引入边界层概念的意义 （3）边界层微分方程组 （1）concept of boundary layer （2）main characteristics of boundary layer and significance of boundary layer concept （3）differential equations of boundary layer	1
5.4	外掠平板层流分析解及比拟理论 analytical solution for the laminar flow over a flat plate and analogy theory	（1）流体外掠等温平壁的对流换热求解 （2）比拟理论的概念 （1）solution of convective heat transfer of flow over an isothermal plate （2）concept of analogy theory	1
5.5	相似原理及量纲分析 similarity principle and dimensional analysis	（1）相似原理是用实验方法求解对流换热问题的重要工具 （2）对流换热常用准则数及物理意义 （1）similarity principle—an important tool for experimental study of convective heat transfer problem （2）common dimensionless numbers in convective heat transfer and their physical significance	1

续表

章节序号 chapter number	章节名称 chapters	知识点 key points	课时 class hour
5.6	相似原理的应用 application of similarity principle	（1）如何安排实验 （2）如何确定待测物理量 （3）如何整理实验数据 （1）how to arrange experiment （2）how to select parameters to be measured （3）how to collate experimental data	1

第六章 单相对流传热的实验关联式（experimental correlations of single-phase convective heat transfer）

章节序号 chapter number	章节名称 chapters	知识点 key points	课时 class hour
6.1	内部强制对流换热的实验关联式 experimental correlations of internal forced convective heat transfer	（1）管槽内强制对流换热的特征 （2）管内湍流对流换热实验关联式 （3）管内层流对流换热实验关联式 （1）characteristics of forced convective heat transfer in tubes and ducts （2）experimental correlations formulas of turbulent convective heat transfer in a tube （3）experimental correlations formulas of laminar convective heat transfer in a tube	1.5
6.2	外部强制对流换热的实验关联式 experimental correlations of external forced convective heat transfer	（1）外掠单管对流换热 （2）外掠管束对流换热 （1）convective heat transfer across single tube （2）convective heat transfer across tube bundle	1
6.3	自然对流换热的实验关联式 experimental correlations of natural convective heat transfer	（1）自然对流流动与换热的特点 （2）大空间自然对流换热实验关联式 （3）有限空间自然对流实验关联式 （1）characteristics of natural convective flow and heat transfer （2）experimental correlations of natural convective heat transfer in large space （3）experimental correlations of natural convective in confined space	1

续表

章节序号 chapter number	章节名称 chapters	知识点 key points	课时 class hour
6.4	强化单相对流传热的技术、机理及性能评价 techniques, mechanism and performance evaluation of the enhancement of single phase convective heat transfer	（1）强化单相对流传热的机理 （2）强化单相对流传热技术 （3）强化单相对流传热技术的性能评价 （1）mechanism of single-phase convective heat transfer enhancement （2）techniques of single-phase convective heat transfer enhancement （3）performance evaluation of single-phase convective heat transfer enhancement	1
6.5	单相对流换热在储能工程中的应用简介 applications of single phase convective heat transfer in energy storage engineering	（1）储能过程涉及的单相对流换热简介 （2）多孔介质中单相对流换热简介 （1）single-phase convective heat transfer in energy storage engineering （2）introduction to single-phase convective heat transfer in porous media	1.5

第七章 相变对流换热的计算（phase change heat transfer calculations）

章节序号 chapter number	章节名称 chapters	知识点 key points	课时 class hour
7.1	凝结换热的模式 model of condensation heat transfer	（1）凝结换热产生条件 （2）珠状凝结 （3）膜状凝结 （1）conditions for the onset of condensation heat transfer （2）dropwise condensation （3）film condensation	0.5
7.2	膜状凝结换热分析解及计算关联式 analytical solution and calculation correlation formula of film condensation heat transfer	（1）努塞特分析解条件及基本步骤 （2）膜状凝结工程计算的流态判据 （3）膜状凝结计算关联式 （1）conditions and basic steps of Nusselt analytical solution （2）criterion for film condensation flow regime in engineering calculation （3）correlations for film condensation	1

续表

章节序号 chapter number	章节名称 chapters	知识点 key points	课时 class hour
7.3	影响膜状凝结换热的因素及强化 factors affecting heat transfer of film condensation and enhancement	（1）影响膜状凝结的因素 （2）凝结换热的强化 （1）factors affecting film-wise condensation （2）enhancement of condensation heat transfer	0.5
7.4	沸腾换热的模式 regimes of boiling heat transfer	（1）沸腾换热的定义及分类 （2）沸腾换热的特点 （3）大容器饱和沸腾曲线 （4）汽泡动力学简介 （1）definition and classification of boiling heat transfer （2）characteristics of boiling heat transfer （3）saturated pool boiling curve （4）introduction of bubble dynamics	0.5
7.5	大容器沸腾换热的实验关联式 experimental correlation formula for pool boiling heat transfer	（1）大容器饱和沸腾的几个区域 （2）核态沸腾换热关联式 （3）临界热流密度的概念及其计算 （4）膜态沸腾换热关联式 （1）different regimes of saturated pool boiling heat transfer （2）correlations for saturation nucleate pool boiling heat transfer （3）concept of critical heat flux and its determination （4）correlation of film boiling heat transfer	1
7.6	沸腾换热的影响因素及强化和热管简介 factors affecting boiling heat transfer and enhancement, and introduction to heat pipe	（1）影响沸腾换热的因素 （2）沸腾换热的强化 （3）热管简介 （1）factors affecting boiling heat transfer （2）enhancement of boiling heat transfer （3）introduction to heat pipe	0.5
7.7	固液及液固相变传热 solid-liquid and liquid-solid phase change heat transfer	（1）固液及液固相变的基本特点 （2）固液及液固相变过程分析 （3）固液相变储能及传热强化 （1）basic features of solid-liquid and liquid-solid phase change （2）theoretical analysis of solid-liquid and liquid-solid phase change process （3）energy storage with solid-liquid phase change and heat transfer intensification	1

第八章 热辐射基本定律和物体的辐射特性（basic laws of thermal radiation and radiative characteristics of materials）

章节序号 chapter number	章节名称 chapters	知识点 key points	课时 class hour
8.1	热辐射现象的基本概念 basic concepts of thermal radiation	（1）辐射传热的本质 （2）辐射传热的特点 （3）吸收比、反射比和穿透比 （4）黑体的定义 （1）essence of thermal radiation （2）characteristics of thermal radiation heat transfer （3）absorptivity, reflectivity and transmissivity （4）definition of black body	1
8.2	黑体热辐射的基本定律 basic laws of black body radiation	（1）斯特藩-玻尔兹曼定律 （2）普朗克定律 （3）兰贝特定律 （1）Stefan-Boltzmann law （2）Planck law （3）Lambert law	1.5
8.3	固体和液体的辐射特性 characteristics of solid and liquid radiation	（1）固体和液体的光谱辐射率 （2）定向辐射强度 （3）影响发射率的因素 （4）液体辐射的特点 （1）solid and liquid spectra-emissive power （2）directional radiation intensity （3）factors affecting emissivity （4）characteristics of liquid radiation	1.5
8.4	气体的辐射特性 characteristics of gas radiation	（1）气体辐射对波长的选择性 （2）气体辐射性质计算简介 （1）spectral selectivity of gas radiation （2）introduction to calculation of gas radiative parameters	1
8.5	实际物体对辐射能的吸收与辐射的关系 relations between absorption and radiation of real bodies	（1）基尔霍夫定律 （2）灰体的概念 （1）Kirchhoff's law （2）concept of grey body	1

第九章　辐射传热的计算（calculation of radiative heat transfer）

章节序号 chapter number	章节名称 chapters	知识点 key points	课时 class hour
9.1	辐射传热的角系数 angle factor of radiative heat transfer	（1）角系数的定义 （2）角系数的性质 （3）角系数的计算方法 （1）definition of angle factor （2）basic features of angle factor （3）determination of angle factor	1.5
9.2	固体表面间辐射传热的计算 calculation of radiative heat transfer between solid surfaces	（1）计算辐射传热的封闭腔概念 （2）有效辐射概念 （3）被透热介质隔开的固体表面间的辐射传热计算方法 （1）enclosure concept of radiative heat transfer calculation （2）concept of effective radiation （3）radiative heat transfer calculation between surface separated by non-radiative medium	1.5
9.3	太阳能利用中的辐射传热 radiative heat transfer in usage of solar energy	（1）太阳辐射的特点 （2）太阳常数 （3）太阳能集热器的工作原理 （1）characteristics of solar radiation （2）solar constant （3）working principle of solar collectors	1.5
9.4	辐射传热的控制及其在储能技术中的应用 control of radiative heat transfer and its application in energy storage engineering	（1）控制物体表面辐射传热的方法 （2）遮热罩及抽气式热电偶 （3）聚光式太阳能热发电站中的储能技术 （1）methods for controlling radiative heat transfer （2）radiative shields and thermal couples with gas extraction （3）energy storage technology in concentrated solar thermal power plant and associated	1.5

第十章 传热过程分析与换热器的热计算（analysis of heat transfer process and thermal calculation of heat exchanger）

章节序号 chapter number	章节名称 chapters	知识点 key points	课时 class hour
10.1	传热过程的分析与计算 analysis and calculation of heat transfer process	（1）传热过程 （2）通过平壁和圆筒壁的传热过程 （3）通过肋壁的传热 （4）临界绝缘直径 （1）heat transfer process （2）heat transfer process in plane and cylinder wall （3）heat transfer process in fins （4）critical insulation diameter	1
10.2	换热器的类型 types of heat exchanger	（1）间壁式换热器 （2）蓄热式换热器 （3）混合式换热器 （1）recuperative heat exchanger （2）regenerative heat exchanger （3）direct contact heat exchanger	1
10.3	换热器平均温差的计算 calculation of mean temperature difference of heat exchanger	（1）顺流与逆流 （2）对数平均温差推导的条件 （3）冷热流体的布置对温差的影响 （1）parallel flow and counter flow （2）condition for derivation of the logarithm mean temperature difference （3）effects of hot and cold flow arrangement on the temperature difference	1
10.4	间壁式换热器的热设计 thermal design of recuperative heat exchanger	（1）设计计算和校核计算 （2）换热器热计算的基本公式 （3）对数平均温差法的计算步骤 （1）design calculation and verfication calculation （2）basic equations for thermal design of heat exchanger （3）calculation procedure for logarithm mean temperature difference method	1.5
10.5	传热控制的主要原理及技术 main principle and technology of controlling heat transfer	（1）强化传热过程的原则 （2）强化传热过程的主要技术 （1）basic principle for enhancing overall heat transfer process （2）major techniques for enhancing heat transfer	1.5

第十一章　质交换（mass transfer）

章节序号 chapter number	章节名称 chapters	知识点 key points	课时 class hour
11.1	质扩散过程与菲克定律 mass diffusion process and Fick's law	（1）混合物浓度 （2）菲克定律 （3）两种典型质扩散过程 （1）concentrations of mixture （2）Fick's law （3）two typical mass diffusion processes	1
11.2	对流传质及表面传质系数 convective mass transfer and surface mass transfer coefficient	（1）对流传质系数 （2）对流传热与对流传质的类比 （3）对流传质计算关联式 （1）convective mass transfer coefficient （2）analogy between convective mass transfer and convective heat transfer （3）correlations of convective mass transfer coefficient	1
11.3	吸附过程及其在储能技术中的应用 adsorption process and its application in energy storage	（1）物理吸附过程 （2）吸附等值线 （3）吸附过程在低品位热能存储中的应用 （1）physical adsorption process （2）adsorption isotherms （3）application of adsorption in storage of low-grade thermal energy	1

5.2.5　上机实验环节（Experiments on computers）

序号 number	实验内容 experiment content	知识点 key points	课时 class hour
1	墙角导热数值计算（数值仿真） numerical simulation of heat conduction in the corner of a wall	用数值计算的方法求出墙角导热的温度场 to calculate temperature field in the corner of a wall with the numerical method	4
2	肋片导热 heat conduction in fin	用数值计算的方法求解肋片导热过程的温度分布 to calculate the temperature field of fin with the numerical method	4

5.3 "热质储能技术及应用(含实验)"教学大纲

课程名称：热质储能技术及应用（含实验）

Course Title：Heat and Mass Energy Storage Technology and Application（with Experiment）

先修课程：储能原理Ⅱ，传热传质学，储能材料工程，储能热流基础（含实验）

Prerequisites：Energy Storage Principle Ⅱ, Heat and Mass Transfer, Energy Storage Material Engineering, Fundamentals of Theamal-Fluid Science in Energy Storage (with Experiment)

学分：3

Credits：3

5.3.1 课程目的和基本内容（Course Objectives and Basic Contents）

本课程是储能科学与工程专业的一门专业必修课。

通过本课程的学习，不仅使学生掌握热质储能技术的基本原理、工作特性和发展动态，同时也培养学生分析和解决具体储能问题的动手能力，为以后深入学习热质储能领域中的相关课程以及从事相关工作打下坚实的基础。热质储能技术及应用主要介绍热质储能技术的基础理论、过程性能分析、系统设计及经济性评价等，涉及的储能技术包括蓄热式换热器、蓄热型热泵、太阳能蓄热技术、蓄冷空调系统、储氢技术等。

热质储能技术及应用通过演示实验、实验设备操作、实验结果分析及讨论等培养过程，使学生对储能系统的构成和工作原理形成直观认识，培养学生的动手能力和分析问题、解决问题的综合能力。

This course is a professional course for undergraduates majoring in energy storage science and engineering.

This course not only enables students to acquire the basic theories, working characteristics and development trends of heat and mass energy storage; but also develops students' ability to analyze and solve energy storage problems and lays a solid foundation for future study and research in heat and mass energy storage. Heat and mass energy storage technology and application mainly introduces the theoretical fundamentals, performance analysis, system design and economic analysis of heat and mass energy storage system. The energy storage technologies involved in this course include regenerative heat exchangers, regenerative heat pump, solar heat storage technology, cool storage air conditioning systems, and hydrogen storage technology, etc.

Through the process of demonstrative experiment, experimental equipment operation, experimental results analysis and discussion in heat and mass energy storage technology and application, students can have an intuitive understanding of the composition, working principle of energy storage system, and cultivate their hands-on ability and comprehensive ability of analyzing and solving problems.

5.3.2 课程基本情况（Basic Information of the Course）

课程名称	热质储能技术及应用（含实验） Heat and Mass Energy Storage Technology and Application（with Experiment）										
开课时间	一年级		二年级		三年级		四年级		总学分	3	总学时	56
	秋	春	秋	春	秋	春	秋	春				
课程定位	储能科学与工程专业本科生热质储能课程群必修课											
授课学时分配	课堂讲授40学时+实验16学时											
先修课程	储能原理II,储能热流基础（含实验）,传热传质学,储能材料工程											
后续课程	先进热力系统技术及仿真,可再生能源及其发电技术											
教学方式	课堂教学,实验与综述报告,讨论,平时作业											
考核方式	课程结束笔试成绩占70%,实验报告占20%,平时成绩占10%											
参考教材	[1] 崔海亭,杨锋. 蓄热技术及其应用. 北京:化学工业出版社,2004. [2] 蔡颖,许剑轶,胡锋,等. 储氢技术与材料. 北京:化学工业出版社,2018.											
参考资料	[1] 沈维道,童钧耕. 工程热力学. 5版. 北京:高等教育出版社,2016. [2] 陶文铨. 传热学. 5版.北京:高等教育出版社,2019. [3] 何雅玲. 工程热力学精要解析.西安:西安交通大学出版社,2014. [4] Dicks A L, Rand D A J. Fuel Cell Systems Explained.Hoboken: John Wiley & Sons, Inc., 2018.											

5.3.3 教学目的和基本要求（Teaching Objectives and Basic Requirements）

（1）使学生深入了解热质储能技术的基本概念、工作原理和应用现状，建立热质储能技术的基础知识体系；

（2）培养学生掌握不同热质储能技术的性能分析方法和适用范围，具体包括蓄热式换热器、蓄热型热泵、太阳能蓄热技术、蓄冷空调系统、储氢技术等；

（3）培养学生针对具体的热质储能问题，运用基础理论、技术和方法解决实际问题的能力，培养科学素养和系统思维能力，为将来从事相关科研工作铺垫坚实基础。

（1）Make students understand the basic concept, working principle and application status of the heat and mass energy storage technology, and establish the basic knowledge system of the heat and mass energy storage.

（2）Cultivate students to master the performance analysis methods and application scope of various heat and mass energy storage technologies, including regenerative heat exchangers, regenerative heat pump, solar heat storage technology, cool storage air conditioning systems, and hydrogen storage technology.

（3）Cultivate students to obtain the ability of solving practical problems by using basic theories, technologies and methods according to specific heat and mass energy storage problems, and have scientific literacy and systematic thinking ability, so as to lay a solid foundation for related scientific research work in the future.

5.3.4 课程大纲和知识点（Syllabus and Key Points）

第一章 绪论（introduction）

章节序号 chapter number	章节名称 chapters	知识点 key points	课时 class hour
1.1	热质储能技术简介 introduction of heat and mass energy storage technology	（1）热质储能技术的基本概念 （2）热质储能技术的特点 （1）concept of heat and mass energy storage technology （2）characteristics of heat and mass energy storage technology	1
1.2	热质储能技术的应用背景与发展历程 application background and development of heat and mass energy storage technology	（1）应用背景 （2）发展历程 （1）application background （2）development history	1
1.3	热质储能技术分类 categories of heat and mass energy storage technology	（1）储热技术 （2）气/液储能技术 （1）heat storage technology （2）gas / liquid energy storage technology	2

第二章 换热器（heat exchangers）

章节序号 chapter number	章节名称 chapters	知识点 key points	课时 class hour
2.1	换热器简介 introduction of heat exchangers	（1）换热器的分类 （2）换热器的结构 （3）换热器的设计方法 （1）types of heat exchangers （2）structures of heat exchangers （3）design methodology of heat exchangers	2
2.2	蓄热式换热器 regenerative heat exchangers	（1）蓄热式换热器的分类 （2）蓄热式换热器的结构和工作原理 （3）蓄热式换热器的设计方法 （1）types of regenerative heat exchangers （2）structures and working principles of regenerative heat exchangers （3）design methodology of regenerative heat exchangers	2

第三章 热质储能容器（containers for heat and mass energy storage）

章节序号 chapter number	章节名称 chapters	知识点 key points	课时 class hour
3.1	蓄热器 heat accumulators	（1）蓄热器的工作原理及结构 （2）蓄热器的热工特性 （3）蓄热器的设计计算 （4）蓄热器的经济性分析 （1）working principles and structures of heat accumulators （2）thermodynamic properties of heat accumulators （3）design methodology of heat accumulators （4）economic analysis of heat accumulators	3
3.2	压力容器 pressure vessels	（1）压力容器的分类 （2）压力容器的发展状况 （3）压力容器的经济性分析 （4）压力容器的应用 （1）types of pressure vessels （2）development of pressure vessels （3）economic analysis of pressure vessels （4）applications of pressure vessels	3

第四章 蓄热型热泵（thermal energy storage heat pumps）

章节序号 chapter number	章节名称 chapters	知识点 key points	课时 class hour
4.1	热泵简介 introduction of heat pumps	（1）热泵的分类 （2）热泵的工作原理 （1）types of heat pumps （2）working principles of heat pumps	3
4.2	蓄热型热泵 regenerative heat pumps	（1）蓄热型热泵的工作原理 （2）蓄热型热泵的介质选择 （3）蓄热型热泵的循环计算 （4）蓄热型热泵的应用 （1）working principles of regenerative heat pumps （2）working fluid selection of regenerative heat pumps （3）cycle calculation of regenerative heat pumps （4）applications of regenerative heat pumps	4

第五章 热质储能技术在太阳能领域的应用（application of heat and mass energy storage technology in solar energy）

章节序号 chapter number	章节名称 chapters	知识点 key points	课时 class hour
5.1	太阳能蓄热技术 solar heat storage technology	（1）太阳能蓄热技术介绍 （2）蓄热技术的分类 （3）工程实例 （1）introduction of solar heat storage technology （2）types of heat storage technology （3）engineering applications	3
5.2	太阳能热发电技术 solar thermal electricity technology	（1）太阳能热发电技术概述 （2）蒸汽动力循环 （3）超临界二氧化碳循环 （4）蓄热型太阳能热发电技术 （5）工程实例 （1）introduction of solar thermal electricity technology （2）steam power cycle （3）supercritical carbon dioxide cycle （4）regenerative solar thermal electricity technology with thermal energy storages （5）engineering applications	4

第六章 热质储能技术在空调领域的应用（application of heat and mass energy storage technology in air conditionings）

章节序号 chapter number	章节名称 chapters	知识点 key points	课时 class hour
6.1	蓄热空调系统 heat storage air conditioning systems	（1）工作原理 （2）系统设计图 （3）设备结构 （4）工程实例 （1）working principles （2）system designs （3）equipment structures （4）engineering applications	3
6.2	蓄冷空调系统 cool storage air conditioning systems	（1）系统简介 （2）分类与基本特性 （3）系统设计 （4）工程实例 （1）introduction of cool storage air conditioning systems	3

续表

章节序号 chapter number	章节名称 chapters	知识点 key points	课时 class hour
6.2	蓄冷空调系统 cool storage air conditioning systems	（2）types and basic characteristics （3）system designs （4）engineering applications	3

第七章　热质储能技术在交通运输领域的应用（application of heat and mass energy storage technology in transportations）

章节序号 chapter number	章节名称 chapters	知识点 key points	课时 class hour
7.1	储氢技术 hydrogen storage technology	（1）氢能概述 （2）气态储氢 （3）液态储氢 （4）固态储氢 （5）发展趋势 （1）introduction of hydrogen storage technology （2）gaseous hydrogen storage （3）liquid hydrogen storage （4）solid hydrogen storage （5）development of hydrogen storage technology	3
7.2	氢燃料电池技术 hydrogen fuel cell technology	（1）工作原理 （2）热力学与动力学特性 （3）燃料电池系统 （4）工程实例 （1）operating principles （2）thermodynamics and kinetics （3）fuel cell systems （4）engineering applications	3

5.3.5　实验环节（Experiments）

序号 number	实验内容 experiment content	知识点 key points	课时 class hour
1	蒸汽动力循环实验 steam power cycle experiment	（1）蒸汽动力循环虚拟演示实验 （2）蒸汽动力循环系统设计 （3）蒸汽动力循环参数计算	4

续表

序号 number	实验内容 experiment content	知识点 key points	课时 class hour
1	蒸汽动力循环实验 steam power cycle experiment	（4）实验设备选配 （5）系统参数测试及数据分析 （6）蒸汽动力循环系统性能讨论 （1）virtual demonstrative experiment of steam power cycle （2）system design of steam power cycle （3）parameters calculation of steam power cycle （4）selection of experiment device （5）system parameter test and data analysis （6）performance discussion of steam power cycle	4
2	制冷循环实验 refrigeration cycle experiment	（1）制冷循环虚拟演示实验 （2）制冷循环系统设计 （3）制冷循环参数计算 （4）实验设备选配 （5）系统参数测试及数据分析 （6）制冷循环系统性能讨论 （1）virtual demonstrative experiment of refrigeration cycle （2）system design of refrigeration cycle （3）parameter calculation of refrigeration cycle （4）selection of experiment device （5）system parameter test and data analysis （6）performance discussion of refrigeration cycle	4
3	二氧化碳临界状态观测及$p\text{-}v\text{-}T$关系测定实验 observation of critical state of carbon dioxide and $p\text{-}v\text{-}T$ relation experiment testing	（1）实验原理介绍 （2）实验装置了解 （3）实验测试与分析 （4）实验报告撰写 （1）introduction of experimental principle （2）familiar with the experimental device （3）experimental test and data analysis （4）experimental report	4
4	换热器综合实验 comprehensive experiment of heat exchanger	（1）换热器综合实验原理介绍 （2）实验装置了解 （3）实验测试与分析 （4）实验报告撰写 （1）introduction of experimental principle for heat exchanger （2）familiar with the experimental device （3）experimental test and data analysis （4）experimental report	4

5.4　"先进热力系统技术及仿真"教学大纲

课程名称：先进热力系统技术及仿真

Course Title：Principle and Simulation of Advanced Thermodynamic Systems

先修课程：储能热流基础（含实验）

Prerequisites：Fundamentals of Thermal-fluid Science in Energy Storage（with Experiments）

学分：2

Credits：2

5.4.1　课程目的和基本内容（Course Objectives and Basic Contents）

本课程是储能科学与工程专业的一门专业选修课。课程主要目的如下：

（1）从热力系统角度出发，对储能及用能过程中的多种热力系统进行介绍，包括热功转换系统、热泵系统、制冷系统、多联产系统等。对上述系统的基本原理进行介绍，着重介绍基本原理和应用现状，重视工程案例分析，着重介绍上述过程的工程概念。

（2）针对上述系统，在建立各个部件模型的基础上，建立系统模型。通过模拟仿真，可以定量了解系统的运行特征并进行相应的分析。

（3）对储能及用能过程中的多种热力系统进行计算机仿真，对系统的模拟方式、模拟工具进行介绍，目的在于让学生掌握系统模拟的方法，了解编程建模的基本思想，使学生掌握在工作中能够直接解决实际问题的能力。

This course is an elective course for undergraduates majoring in energy storage science and engineering. The main objectives of the course are:

（1）From the perspective of thermal systems, various thermal systems in the process of energy storage and energy usage are introduced, including heat to power conversion systems, heat pump systems, refrigeration systems, multi-generation systems, and so on. The basic concepts are introduced focused on the basic principles and application status. Engineering case studies are carried out and emphasized. The introduction of engineering concept in the above process is also emphasized.

（2）For the above-mentioned systems, models of each component and the system are introduced. The performance characteristics of the systems are acquired and analyzed according to the simulation results.

（3）Simulation of various thermal systems for energy storage and utilization are carried out. The simulation methods and simulation tools are introduced. The purpose is to help students to master the methods of system simulation, understand the basic ideas of modeling, and obtain the ability to solve practical problems.

5.4.2 课程基本情况（Basic Information of the Course）

课程名称	先进热力系统技术及仿真 Principle and Simulation of Advanced Thermodynamic Systems							
开课时间	一年级		二年级		三年级	四年级	总学分	2	总学时	32
	秋	春	秋	春	秋	春	秋	春				
课程定位	储能科学与工程专业本科生热质储能课程群选修课							
授课学时分配	课堂讲授32学时							
先修课程	储能热流基础（含实验）							
后续课程	热质储能综合实验							
教学方式	课堂教学，大作业，平时作业							
考核方式	课程结束笔试成绩占70%，平时成绩占30%							
参考教材	无							
参考资料	[1] 张会生，周登极，翁史烈. 热力系统建模与仿真技术.上海：上海交通大学出版社，2018. [2] Gicquel R. Energy systems: a new approach to engineering thermodynamics. Boca Raton: CRC Press, 2011. [3] Wu C. Thermodynamics and heat powered cycles: a cognitive engineering approach. New York: Nova Science Publishers, 2007.							

5.4.3 教学目的和基本要求（Teaching Objectives and Basic Requirements）

（1）掌握储能及用能过程中的多种热力系统的基本原理；

（2）熟练掌握上述系统的功能及性能计算方法；

（3）掌握工业过程中换热器、汽轮机、泵等核心部件的原理及模拟方法；

（4）掌握工业过程中膨胀、压缩、传热、燃烧等热力学过程的模拟方法；

（5）掌握热力系统计算机模拟的基本思路，掌握系统模拟方法，了解自编程建模的基本思想；

（6）能够对既定热力系统进行建模仿真，模拟不同参数下的系统性能。

（1）Master the basic principles of various thermal systems in the process of energy storage and energy utilization.

（2）Get familiar with the functions and performance calculation methods of the above systems.

（3）Master the principles and simulation methods of core components such as heat exchangers, steam turbines, and pumps in industrial processes.

（4）Master the simulation methods of expansion, compression, heat transfer, combustion, and other thermodynamic processes in industrial processes.

（5）Master the basic ideas of computer simulation for thermal systems. Master system simulation methods, and understand the basic ideas of self-programming modeling.

（6）Be able to model and simulate the established thermal system and simulate the system performance with different parameters.

5.4.4　课程大纲和知识点（Syllabus and Key Points）

第一章　绪论（introduction）

章节序号 chapter number	章节名称 chapters	知识点 key points	课时 class hour
1.1	可再生能源现状 current status of renewable energy	（1）国家能源现状 （2）可再生能源利用现状 （1）current status of national energy （2）current status of renewable energy utilization	2
1.2	可再生能源的多样性 diversity of renewable energy	（1）可再生能源利用中的问题 （2）能量的品位 （1）problems in the utilization of renewable energy （2）grade of energy	
1.3	可再生能源的利用及储存方式 utilization and storage methods for renewable energy	能量利用及储存方式综述 overview of energy utilization and storage methods	

第二章　热力学的分析方法（thermodynamic analysis methods）

章节序号 chapter number	章节名称 chapters	知识点 key points	课时 class hour
2.1	熵和焓 entropy and enthalpy	（1）熵的工业意义 （2）焓的工业意义 （3）等熵效率和卡诺循环 （4）热、功 （5）可用能的概念 （1）industrial significance of entropy （2）industrial significance of enthalpy （3）isentropic efficiency and Carnot cycle （4）heat and work （5）concept of energy	2

续表

章节序号 chapter number	章节名称 chapters	知识点 key points	课时 class hour
2.2	㶲分析方法 exergetic analysis method	（1）㶲的含义 （2）㶲损失 （3）系统的㶲效率 （1）meaning of exergy （2）exergetic loss （3）exergetic efficiency of the system	2
2.3	热力系统的基本概念 basic concept of thermal system	（1）换热器热负荷计算 （2）泵功计算 （3）汽轮机输出功计算 （4）系统热效率、㶲效率计算 （1）heat load calculation of heat exchanger （2）calculation of pump power （3）calculation of output power of steam turbine （4）calculations of system thermal efficiency and exergetic efficiency	2

第三章 热力过程计算机模拟（computer simulation of thermodynamic process）

章节序号 chapter number	章节名称 chapters	知识点 key points	课时 class hour
3.1	建模的物理意义与现实意义 physical and practical significance of modeling	（1）建模的意义 （2）建模的步骤 （3）热力过程商业软件建模及自编程建模的实现方法 （1）significance of modeling （2）modeling procedure （3）simulation of the thermodynamic process via commercial software and self-programming	2
3.2	部件模型建立 development of the component model		
3.3	热力过程模拟在商业软件中的实现 simulation of the thermodynamic process via commercial software		2
3.4	热力过程模拟的自编程实现 simulation of the thermodynamic process via self-programming		

第四章　热功转换系统（heat-power system）

章节序号 chapter number	章节名称 chapters	知识点 key points	课时 class hour
4.1	朗肯循环 Rankine cycle	（1）朗肯循环的基本原理 （2）朗肯循环的性能计算 （3）朗肯循环的应用 （1）basic principle of Rankine cycle （2）performance calculation of Rankine cycle （3）application of Rankine cycle	2
4.2	布雷顿循环 Brayton cycle	（1）布雷顿循环的基本原理 （2）布雷顿循环的性能计算 （3）布雷顿循环的应用 （1）basic principle of Brayton cycle （2）performance calculation of Brayton cycle （3）application of Brayton cycle	2
4.3	有机朗肯循环 organic Rankine cycle	（1）有机朗肯循环的基本原理 （2）有机朗肯循环的性能计算 （3）有机朗肯循环的应用 （1）basic principle of organic Rankine cycle （2）performance calculation of organic Rankine cycle （3）application of organic Rankine cycle	2
4.4	卡林那循环 Kalina cycle	（1）卡林那循环的基本原理 （2）卡林那循环的性能计算 （3）卡林那循环的应用 （1）basic principle of Kalina cycle （2）performance calculation of Kalina cycle （3）application of Kalina cycle	2

第五章　热泵及制冷系统 (heat pump and refrigeration system)

章节序号 chapter number	章节名称 chapters	知识点 key points	课时 class hour
5.1	蒸汽压缩式制冷/热泵 vapor compression refrigeration/heat pump	（1）蒸汽压缩式制冷/热泵基本原理 （2）蒸汽压缩式制冷/热泵的性能计算 （3）蒸汽压缩式制冷/热泵的应用 （1）basic principles of vapor compression refrigeration/heat pump （2）performance calculation of vapor compression refrigeration/heat pump （3）application of vapor compression refrigeration/heat pump	2

续表

章节序号 chapter number	章节名称 chapters	知识点 key points	课时 class hour
5.2	吸收式制冷/热泵 absorption refrigeration/ heat pump	（1）吸收式制冷/热泵基本原理 （2）吸收式制冷/热泵的性能计算 （3）吸收式制冷/热泵的应用 （1）basic principles of absorption refrigeration/heat pump （2）performance calculation of absorption refrigeration/heat pump （3）application of absorption refrigeration/heat pump	2

第六章 多联产技术（multi-generation systems）

章节序号 chapter number	章节名称 chapters	知识点 key points	课时 class hour
6.1	热电联产技术 combined heat and power （CHP）	（1）典型热电联产系统构建 （2）典型热电联产系统性能计算 （3）热电联产技术的应用 （1）construction of a typical CHP system （2）performance calculation of a typical CHP system （3）application of CHP technology	2
6.2	冷热电三联供技术 combined cool, heat and power（CCHP）	（1）典型冷热电三联供系统构建 （2）典型冷热电三联供系统性能计算 （3）冷热电三联供技术的应用 （1）construction of a typical CCHP system （2）performance calculation of a typical CCHP system （3）application of CCHP technology	2
6.3	分布式能源系统 distributed energy system	（1）分布式能源系统的意义 （2）典型分布式能源系统展示及分析 （1）significance of distributed energy systems （2）demonstration and analysis of typical distributed energy systems	2
6.4	可再生能源耦合储能系统 renewable energy coupled energy storage system	（1）可再生能源耦合储能系统的意义 （2）典型可再生能源耦合储能系统展示及分析 （1）significance of renewable energy coupled energy storage system （2）demonstration and analysis of typical renewable energy coupled energy storage system	2

5.5 "可再生能源及其发电技术"教学大纲

课程名称：可再生能源及其发电技术
Course Title：Renewable Energy and Its Power Generation Technology
先修课程：储能热流基础（含实验），储能原理Ⅰ，储能原理Ⅱ
Prerequisites：Fundamentals of Thermal-Fluid Science in Energy Storage（with Experiments），Energy Storage Principle Ⅰ，Energy Storage Principle Ⅱ
学分：2
Credits：2

5.5.1 课程目的和基本内容（Course Objectives and Basic Contents）

本课程是储能科学与工程专业的一门专业选修课。

通过本课程的学习，使学生掌握可再生能源（太阳能、风能、生物质能等）的特点及其应用情况，掌握可再生能源发电及储能系统的基本构成、工作原理及系统性能分析和评价方法，受到很好的基本技能训练，能正确进行可再生能源发电及其储能系统的计算和分析。

本课程以可再生能源及其发电技术为核心，分别介绍太阳能、生物质能、风能等可再生能源的特点，可再生能源发电技术研究进展，储能技术在可再生能源发电系统中的应用及案例分析，重点介绍可再生能源发电系统的构成、工作原理及性能分析和评价方法。

This course is an elective course for undergraduates majoring in energy storage science and engineering.

Through the study of this course, students can master the characteristics and applications of renewable energy (solar energy, wind energy, biomass energy, etc.), the basic composition, working principle, analysis method of system performance, and evaluation method in terms of the renewable energy power generation and energy storage system, and will be well trained in basic skills to properly calculate and analyze the renewable energy generation and energy storage system.

The course focuses on renewable energy and its power generation technology. The characteristics of solar energy, biomass energy, wind energy, and other renewable energy will be introduced. The research progress of renewable energy power generation technology, the application of energy storage technology in renewable energy power generation system, and the typical application cases will be discussed. The emphasis is on the composition, working principle, and performance analyses and evaluation methods of the renewable energy power generation system.

5.5.2 课程基本情况（Basic Information of the Course）

课程名称	可再生能源及其发电技术 Renewable Energy and Its Power Generation Technology							
开课时间	一年级	二年级	三年级	四年级	总学分	2	总学时	32
	秋　春	秋　春	秋　春	秋　春				
课程定位	储能科学与工程专业本科生热质储能课程群选修课							
授课学时分配	课堂讲授28学时+讨论4学时							
先修课程	储能热流基础（含实验），储能原理I，储能原理II							
后续课程	无							
教学方式	课堂教学，大作业							
考核方式	课程结束笔试成绩占60%，大作业成绩占30%，平时成绩占10%							
参考教材	[1] 姚兴佳. 可再生能源及其发电技术. 北京：科学出版社，2010.							
参考资料	[1] 尹忠东，朱永强. 可再生能源发电技术. 北京：中国水利水电出版社，2010. [2] 何一鸣，钱显毅，刘龙春，等. 可再生能源及其发电技术. 北京：北京交通大学出版社，2013.							

5.5.3 教学目的和基本要求（Teaching Objectives and Basic Requirements）

通过本课程的学习，使学生获得可再生能源发电方面的基本理论、基本知识和基本技能，为后续课程的学习及将来从事相关科研工作打下一定的基础。课程基本要求如下：

（1）掌握可再生能源的概念和内涵；

（2）熟悉太阳能、生物质能、风能等可再生能源的特点；

（3）掌握可再生能源发电系统的基本原理、发展历史和应用现状；

（4）熟悉储能技术在可再生能源发电技术中的应用现状；

（5）掌握可再生能源发电系统性能分析和评价方法。

Through the study of this course, students can master the basic theory, knowledge, and skills of renewable energy and its power generation technology, and lay the foundation for the subsequent study of related courses and related research. Basic requirements are as follows:

（1）Master the concept and connotation of renewable energy.

（2）Be acquainted with the characteristics of solar energy, biomass energy, and wind energy.

（3）Master the basic principles, development history, and application status of renewable energy power generation technology.

（4）Be acquainted with the application status of energy storage in renewable energy power generation technology.

（5）Master the analyses and evaluation methods of renewable energy power generation systems.

5.5.4　课程大纲和知识点（Syllabus and Key Points）

第一章　绪论（introduction）

章节序号 chapter number	章节名称 chapters	知识点 key points	课时 class hour
1.1	能源概述 energy overview	（1）能源的定义 （2）能源的分类 （3）可再生能源的内涵 （1）definition of energy （2）classification of energy （3）connotation of renewable energy	0.5
1.2	能源与环境 energy and environmental	（1）能源现状 （2）能源与环境 （3）可再生能源的应用 （1）energy status （2）energy and environmental （3）application of renewable energy	0.5
1.3	可再生能源发电 renewable energy power generation	（1）太阳能发电技术 （2）生物质能发电技术 （3）风能发电技术 （1）solar power generation technology （2）biomass power generation technology （3）wind power generation technology	1

第二章　太阳能热发电（solar thermal electricity）

章节序号 chapter number	章节名称 chapters	知识点 key points	课时 class hour
2.1	太阳能的基础知识 basic knowledge of solar energy	（1）太阳辐射与太阳能 （2）太阳辐射参数 （3）太阳能资源及其分布 （4）太阳能热发电系统构成 （1）solar radiation and solar energy （2）solar radiation parameters （3）solar energy resource and its distribution （4）system composition of solar thermal electricity	1

续表

章节序号 chapter number	章节名称 chapters	知识点 key points	课时 class hour
2.2	太阳能聚光器的工作原理 working principle of solar concentrator	（1）槽式聚光器 （2）塔式聚光器 （3）碟式聚光器 （4）菲涅尔聚光器 （1）parabolic trough concentrator （2）tower concentrator （3）dish concentrator （4）fresnel concentrator	2
2.3	太阳能吸热器的工作原理 working principle of solar receiver	（1）真空管式吸热器 （2）腔体式吸热器 （3）光热转换原理 （4）热能传递机理 （1）vacuum tube receiver （2）cavity receiver （3）photothermal conversion mechanism （4）heat transfer mechanism	2
2.4	用于太阳能热发电的动力循环性能分析 performance analysis of power cycle used in solar thermal power generation	（1）有机朗肯循环 （2）水蒸气动力循环 （3）气体动力循环 （1）organic Rankine cycle （2）steam power cycle （3）gas power cycle	1

第三章　太阳能光伏发电（solar photovoltaic power generation）

章节序号 chapter number	章节名称 chapters	知识点 key points	课时 class hour
3.1	太阳光伏发电系统 solar photovoltaic power generation system	（1）系统构成 （2）系统分类 （3）光伏电池研究进展 （1）system composition （2）system classification （3）research progress of photovoltaic cells	1
3.2	光伏电池与阵列 photovoltaic cells and arrays	（1）电池工作原理 （2）电池基本特性 （3）电池组件与阵列 （1）working principle	2

章节序号 chapter number	章节名称 chapters	知识点 key points	课时 class hour
3.2	光伏电池与阵列 photovoltaic cells and arrays	（2）basic characteristics （3）cells and arrays	2
3.3	光伏发电系统设计 design of photovoltaic power generation system	（1）容量设计 （2）优化设计 （3）案例分析 （1）capacity design （2）optimal design （3）case analysis	2
3.4	光伏发电的应用与发展 application and development of photovoltaic power generation	（1）光伏材料的研究进展 （2）光伏发电系统研究进展 （3）未来展望 （1）research progress of photovoltaic materials （2）research progress of photovoltaic power generation system （3）future prospects	1

第四章　风力发电（wind power generation）

章节序号 chapter number	章节名称 chapters	知识点 key points	课时 class hour
4.1	风力发电机组构成 composition of wind turbine	（1）风能特性 （2）机组构成 （3）运行方式 （1）wind energy characteristics （2）composition of wind turbine （3）operation mode	1
4.2	风力发电的理论基础 theoretical basis of wind power generation	（1）升力与阻力 （2）工作原理 （3）风轮功率 （4）系统效率与有效功率 （1）lift and drag （2）working principle （3）wind turbine power （4）system efficiency and effective power	2
4.3	风力发电机简介 brief introduction of wind generator set	（1）风力发电机分类 （2）笼型感应发电机 （3）同步发电机 （4）绕线转子感应发电机	1

续表

章节序号 chapter number	章节名称 chapters	知识点 key points	课时 class hour
4.3	风力发电机简介 brief introduction of wind generator set	（1）classification of wind generator set （2）cage induction generator （3）synchronous generator （4）wound rotor induction generator	1
4.4	风力发电系统控制策略 control strategy of wind power generation system	（1）控制系统的构成 （2）定桨距失速控制 （3）变桨距控制 （4）偏航控制 （1）composition of control system （2）fixed pitch stall control （3）variable pitch control （4）yaw control	2

第五章 生物质能发电（biomass power generation）

章节序号 chapter number	章节名称 chapters	知识点 key points	课时 class hour
5.1	生物质能资源及统计 biomass energy resources and statistics	（1）生物质与生物质能 （2）生物质的热值 （3）生物质资源量统计方法 （4）生物质能发电途径 （1）biomass and biomass energy （2）calorific value of biomass （3）statistical methods of biomass resources （4）ways of biomass power generation	1
5.2	生物质直接燃烧发电 biomass direct combustion power generation	（1）系统构成 （2）锅炉工作原理 （3）汽轮机工作原理 （4）典型案例分析 （1）system composition （2）working principle of boiler （3）working principle of steam turbine （4）typical case analysis	2
5.3	生物质气化发电 biomass gasification power generation	（1）生物质气化原理 （2）生物质气化炉 （3）生物质燃气的净化 （4）典型案例分析	2

续表

章节序号 chapter number	章节名称 chapters	知识点 key points	课时 class hour
5.3	生物质气化发电 biomass gasification power generation	（1）principle of biomass gasification （2）biomass gasifier （3）purification of biomass gas （4）typical case analysis	2
5.4	生物质发电成本分析 cost analysis of biomass power generation	（1）不变成本分析 （2）可变成本分析 （3）原料成本区域性分析 （1）constant cost analysis （2）variable cost analysis （3）regional analysis of raw material cost	1

第六章 储能技术在可再生能源发电中的应用（application of energy storage technology in renewable energy power generation）

章节序号 chapter number	章节名称 chapters	知识点 key points	课时 class hour
6.1	储热技术在可再生能源发电中的应用 application of heat storage technology in renewable energy power generation	（1）系统构成及工作原理 （2）系统性能分析方法 （3）系统性能评价与优化 （4）典型案例分析 （1）system composition and working principle （2）performance analysis method （3）performance evaluation and optimization （4）typical case analysis	3
6.2	储电技术在可再生能源发电中的应用 application of power storage technology in renewable energy power generation	（1）系统工作原理 （2）系统性能分析 （3）系统性能评价与优化 （4）典型案例分析 （1）system working principle （2）system performance analysis （3）system performance evaluation and optimization （4）typical case analysis	3

5.6 "流体机械原理及其储能应用"教学大纲

课程名称：流体机械原理及其储能应用

Course Title：Principles of Fluid Machinery and Its Application in Energy Storage

先修课程：高等数学Ⅰ，大学物理（含实验），工程力学，储能热流基础（含实验）

Prerequisites：Advanced Mathematics Ⅰ, Physics and Physics Experiments, Engineering Mechanics, Fundamentals of Thermal-Fluid Science in Energy Storage （with Experiments）

学分：2

Credits：2

5.6.1 课程目的和基本内容（Course Objectives and Basic Contents）

本课程是储能科学与工程专业的一门专业选修课。

本课程介绍离心式压缩机基本结构、基本工作原理和性能实验基本技能。灵活运用已学的基础知识解决专业课程中的实际问题，培养工程实践观念。培养学生严谨求实的科学态度、正确的思维方法、独立分析和解决实际问题的能力、初步的创新意识。

本课程的主要内容包括离心式压缩机基本工作原理、压缩机基本结构和各个通流元件的主要作用、气体在压缩机中流动及热力状态变化的基本方程和基本概念、能量损失机理及分析、叶轮和固定元件、压缩机相似理论基础及应用等。课程设有离心压缩机性能实验，测试并绘制出压力、效率、功率、噪声随流量变化的关系曲线。本课程是从事储能科学与工程设计、运行和管理工作的高级人才的必备课程。

This course is an elective course for undergraduates majoring in energy storage science and engineering.

This course introduces the basic structure, basic working principle and basic skills of performance test of centrifugal compressor. This course focuses on applying the basic knowledge flexibly to solve practical problems in professional courses and cultivate the concept of engineering practice. For students, this course cultivate their rigorous and realistic scientific attitude, correct thinking method, independent analysis and solutions for practical problems, and preliminary innovation consciousness.

The course focuses on the basic principle of centrifugal compressor, including the basic structure of compressor and the main role of each flow component, the basic equation and basic concept of gas flow and thermodynamic state changes in the compressor, the mechanism and analysis of energy losses, impeller and fixed elements, the theoretical basis and application of compressor similitude, etc. In this course, compressor performance experiment is set up to test and draw the relation curve of pressure, efficiency, power and noise with the change of flowrate. This course is an improtant course for senior talents engaged in design, operation and management of energy storage science and engineering.

5.6.2 课程基本情况（Basic Information of the Course）

课程名称	流体机械原理及其储能应用 Principles of Fluid Machinery and Its Application in Energy Storage											
开课时间	一年级		二年级		三年级		四年级		总学分	2	总学时	36
	秋	春	秋	春	秋	春	秋	春				
课程定位	储能科学与工程专业本科生热质储能模块选修课											
授课学时分配	课堂讲授32学时+实验4学时											
先修课程	高等数学Ⅰ,大学物理（含实验),工程力学,储能热流基础（含实验）											
后续课程	无											
教学方式	课堂教学,大作业,实验,讨论,平时作业											
考核方式	课程结束笔试成绩占70%,实验报告占20%,平时成绩占10%											
参考教材	[1] 祁大同. 离心压缩机原理.北京:机械工业出版社,2018. [2] 徐忠. 离心压缩机原理.北京:机械工业出版社,1990. [3] 朱报祯,郭涛.离心压缩机.西安:西安交通大学出版社,1989.											
参考资料	无											

5.6.3 教学目的和基本要求（Teaching Objectives and Basic Requirements）

（1）掌握离心式压缩机的基本结构和基本工作原理；
（2）掌握离心式压缩机主要参数对性能的影响；
（3）掌握离心式压缩机热力参数计算的基本方法；
（4）掌握离心式压缩机性能相似换算的基本方法；
（5）培养严谨的学习态度、正确的思维方法、基本的工程观点和初步的创新意识。

（1）Master the basic structure and working principle of centrifugal compressor.

（2）Master the influence of main parameters on performance of centrifugal compressor.

（3）Master the basic method of calculation of thermodynamic parameters in centrifugal compressor.

（4）Master the basic method of similarity conversion of centrifugal compressor performance.

（5）Cultivate rigorous learning attitude, correct thinking method, basic engineering viewpoint and preliminary innovation consciousness.

5.6.4 课程大纲和知识点（Syllabus and Key Points）

第一章 绪论（introduction）

章节序号 chapter number	章节名称 chapters	知识点 key points	课时 class hour
1.1	透平压缩机的分类 classification of turbine compressor	（1）透平压缩机的分类 （2）透平压缩机的应用 （1）classification of turbine compressor （2）application of turbine compressor	1
1.2	透平压缩机的现状、发展趋势及在储能中的应用 state of the art and development trend of turbine compressor and its application in energy storage	（1）透平压缩机技术现状及发展趋势 （2）透平压缩机在储能中的应用 （1）technical status and development trend of turbine compressor （2）application of turbine compressor in energy storage	1

第二章 气体流动的基本概念和基本方程（basic concepts and basic equations of gas flow）

章节序号 chapter number	章节名称 chapters	知识点 key points	课时 class hour
2.1	基本假设和速度三角形 basic assumptions and velocity triangle	（1）基本假设 （2）速度三角形及速度分量 （1）basic assumptions （2）velocity triangle and velocity component	1
2.2	气体流动基本方程 basic equations of gas flow	（1）状态方程 （2）连续性方程 （3）欧拉方程 （4）能量方程 （5）伯努利方程 （6）过程方程 （1）state equation （2）continuity equation （3）Euler equation （4）energy equation （5）Bernoulli's equation （6）process equation	3
2.3	气体压缩过程及压缩功 gas compression process and compression work	（1）气体压缩过程 （2）压缩功 （1）gas compression process （2）compression work	1

章节序号 chapter number	章节名称 chapters	知识点 key points	课时 class hour
2.4	级效率 stage efficiency	（1）级效率 （2）多变效率 （3）等熵效率 （4）等温效率 （1）stage efficiency （2）polytrophic efficiency （3）isentropic efficiency （4）isothermal efficiency	1
2.5	级中气流参数变化及计算 variation and calculation of airflow parameters in the stage	（1）滞止温度、滞止压力 （2）级中气流参数变化及计算 （1）stagnation temperature and stagnation pressure （2）variation and calculation of airflow parameters in the stage	1
2.6	轴向涡流 axial vortex	（1）轴向涡流 （2）周速系数 （1）axial vortex （2）circumferential velocity coefficient	1

第三章　级中能量损失及级性能曲线（stage energy loss and stage performance curve）

章节序号 chapter number	章节名称 chapters	知识点 key points	课时 class hour
3.1	流动损失 flow loss	（1）摩擦损失 （2）分离损失 （3）二次流损失 （4）尾迹损失 （1）friction loss （2）separation loss （3）secondary flow loss （4）wake loss	1
3.2	级性能曲线 stage performance curve	（1）多变能量头与流量的关系 （2）级性能曲线 （1）relationship between polytropic energy head and flow rate （2）stage performance curve	

续表

章节序号 chapter number	章节名称 chapters	知识点 key points	课时 class hour
3.3	雷诺数和马赫数对流动损失的影响 influence of Reynolds number and Mach number on flow loss	（1）雷诺数对流动损失的影响 （2）马赫数对流动损失的影响 （1）influence of Reynolds number on flow loss （2）influence of Mach number on flow loss	1
3.4	泄漏损失和轮阻损失 leakage loss and disc friction loss	（1）泄漏损失 （2）轮阻损失 （1）leakage loss （2）disc friction loss	

第四章　叶轮（impeller）

章节序号 chapter number	章节名称 chapters	知识点 key points	课时 class hour
4.1	叶轮主要结构参数 main structural parameters of impeller	（1）叶轮主要结构参数 （2）几种不同的叶片形式 （1）main structural parameters of impeller （2）several different blade types	1
4.2	不同叶片形式的综合比较 comprehensive comparison of different blade types	（1）反作用度 （2）叶轮效率 （3）不同叶片形式的综合比较 （1）reaction degree （2）impeller efficiency （3）comprehensive comparison of different blade types	1
4.3	周速系数及其计算 circumferential velocity coefficient and its calculation	（1）周速系数 （2）周速系数经验公式 （1）circumferential velocity coefficient （2）empirical formula of circumferential velocity coefficient	1
4.4	叶轮主要结构参数对性能的影响 influence of main parameters of impeller on stage performance	（1）叶轮进口几何参数对性能的影响 （2）叶轮出口几何参数对性能的影响 （3）叶片数对性能的影响 （1）influence of geometric parameters at impeller inlet on stage performance （2）influence of geometric parameters at impeller outlet on stage performance （3）influence of blade number on stage performance	1

第五章　固定原件（stators）

章节序号 chapter number	章节名称 chapters	知识点 key points	课时 class hour
5.1	吸气室 suction chamber	（1）几种典型结构 （2）吸气室内的流动分析 （1）several typical structures of suction chamber （2）flow analysis in suction chamber	1
5.2	扩压器 diffuser	（1）无叶扩压器 （2）叶片扩压器 （3）直壁形扩压器 （1）vaneless diffuser （2）vane diffuser （3）straight wall diffuser	1
5.3	弯道和回流器 bends and returns	（1）弯道 （2）回流器 （1）bends （2）returns	1
5.4	蜗壳 volute	（1）蜗壳结构 （2）蜗壳内流动分析 （3）蜗壳内壁型线 （1）structure of volute （2）flow analysis of volute （3）inner profile of volute	1

第六章　流动相似理论在离心压缩机中的应用（application of flow similarity theory in centrifugal compressor）

章节序号 chapter number	章节名称 chapters	知识点 key points	课时 class hour
6.1	离心压缩机流动相似条件 similitude condition of centrifugal compressor	（1）流动相似准则数 （2）几何相似 （3）运动相似 （4）动力相似 （5）热力相似 （1）flow similarity numbers （2）geometric similarity （3）kinematic similarity （4）dynamic similarity （5）thermodynamic similarity	2

续表

章节序号 chapter number	章节名称 chapters	知识点 key points	课时 class hour
6.2	离心压缩机性能相似换算 performance similarity conversion of centrifugal compressor	（1）离心压缩机性能相似换算 （2）无因次参数及无因次性能曲线 （3）比转速 （1）performance similarity conversion of centrifugal compressor （2）dimensionless parameters and dimensionless performance curves （3）specific rotational speed	1
6.3	离心压缩机相似设计 design of centrifugal compressor based on similitude	（1）离心压缩机级相似设计 （2）有中冷器时机器的模化设计 （1）similarity design of centrifugal compressor stage （2）similarity design for machines with intercoolers	1

第七章 离心压缩机的性能曲线和调节（performance curve and adjustment of centrifugal compressor）

章节序号 chapter number	章节名称 chapters	知识点 key points	课时 class hour
7.1	离心压缩机的性能曲线 performance curve of centrifugal compressor	（1）级数对离心压缩机性能的影响 （2）转速对离心压缩机性能的影响 （1）effect of multistages on the performance of centrifugal compressor （2）effect of rotating speed on performance of centrifugal compressor	1
7.2	离心压缩机与管网联合工作 coupling operation of centrifugal compressor and fluid network	（1）离心压缩机与管网联合工作 （2）离心压缩机串联工作 （3）离心压缩机并联工作 （1）coupling operation of centrifugal compressor and fluid network （2）centrifugal compressor in series （3）centrifugal compressor in parallel	1
7.3	旋转失速与喘振 rotating stall and surge	（1）旋转失速 （2）喘振 （3）喘振控制 （1）stall （2）surge （3）surge control	1

章节序号 chapter number	章节名称 chapters	知识点 key points	课时 class hour
7.4	离心压缩机基本调节方法 adjustment methods of centrifugal compressor	（1）进口节流调节 （2）出口调节 （3）变转速调节 （4）其他调节方法 （1）inlet adjustment methods （2）outlet adjustment methods （3）variable speed regulation （4）other adjustment methods	1

第八章　离心压缩机在压缩空气储能中的应用（application of centrifugal compressor in compressed air energy storage）

章节序号 chapter number	章节名称 chapters	知识点 key points	课时 class hour
8.1	压缩空气储能系统 compressed air energy storage system	（1）压缩空气储能系统 （2）离心压缩机在储能系统中的作用 （1）compressed air energy storage system （2）role of centrifugal compressor in energy storage system	1
8.2	压缩空气储能系统中离心压缩机的运行特征 operation characteristics of centrifugal compressor in compressed air energy storage system	（1）压缩空气储能系统离心压缩机运行特点 （2）压缩空气储能系统离心压缩机的需求与挑战 （1）operation characteristics of centrifugal compressor in compressed air energy storage system （2）demand and challenge of centrifugal compressor in compressed air energy storage system	1
8.3	压缩空气储能系统离心压缩机的选型与维护 selection and maintenance of centrifugal compressor for compressed air energy storage system	（1）压缩空气储能系统离心压缩机选型 （2）压缩空气储能系统离心压缩机的调节与维护 （1）selection of centrifugal compressor for compressed air energy storage system （2）regulation and maintenance of centrifugal compressor in compressed air energy storage system	2

5.6.5 实验环节（Experiments）

序号 number	实验内容 experiment content	知识点 key points	课时 class hour
1	离心压缩机性能实验 performance test of centrifugal compressor	（1）测量参数及测量方法 （2）数据处理及性能换算 （3）实验结果分析 （1）measurement parameters and methods （2）data processing and performance conversion （3）analysis of test results	4

5.7 "氢能储存与应用"教学大纲

课程名称：氢能储存与应用
Course Title：Hydrogen Storage and Application
先修课程：储能热流基础（含实验），储能化学基础（含实验），大学物理（含实验）
Prerequisites：Fundamentals of Thermal-Fluid Science in Energy Storage（with Experiments），General Chemistry for Energy Storage（with Experiments），Physics and Physics Experiments
学分：2
Credits：2

5.7.1 课程目的和基本内容（Course Objectives and Basic Contents）

本课程是储能科学与工程专业的一门专业选修课。

氢能被视为 21 世纪最具发展潜力的清洁能源之一。利用太阳能、风能、生物质能等可再生能源制氢，通过氢燃料电池发电，产物为水，可以缓解化石能源危机，解决环境污染问题，由此形成的零排放氢经济是未来社会一种理想的经济结构模式。过去 10 年，氢的制取、储存、运输、应用技术已取得了长足发展，并成为备受关注的研究热点。通过氢能储存与应用课程的学习，使学生了解国内外氢能的发展现状和未来发展趋势，掌握氢能发展的关键科学与技术问题。

氢能储存与应用课程的主要学习内容包括氢能的开发与利用；制氢技术，包括化石燃料制氢和可再生能源制氢；氢能的储存技术，包括储氢材料与储氢容器；液氢的生产、储存与运输；氢能的应用，主要包括氢燃料电池原理和技术。

This course is an elective course for undergraduates majoring in energy storage science and engineering.

The hydrogen has been considered as one of the most promising clean energies in the 21st century. By utilizing renewable energy such as solar energy, wind power, biomass, etc. to produce hydrogen, and then use hydrogen to generate electricity through hydrogen fuel cell with the product of water, such zero-emission hydrogen energy economy is an ideal economic model, which would greatly relieve both the energy crisis of fossil fuel and the environmental pollution. During the past decade, the production, storage, transportation, and utilization of the hydrogen have received considerable development, and thus the hydrogen economy becomes an attractive research topic. Via the course of hydrogen storage and application, students can recognize the application status and development trend of hydrogen energy, as well as master the key scientific and technological issues in the hydrogen development.

The topics covered in this course include the development and utilization of hydrogen energy, hydrogen production technology from fossil fuel to renewable energy, hydrogen storage technology including materials and vessels, liquid hydrogen technology from production, storage to transportation, hydrogen application including principle and technology of hydrogen fuel cell.

5.7.2　课程基本情况（Course Arrangements）

课程名称	氢能储存与应用 Hydrogen Storage and Application										
开课时间	一年级		二年级		三年级		四年级		总学分	2	总学时	32
	秋	春	秋	春	秋	春	秋	春				
课程定位	储能科学与工程专业本科生热质储能模块选修课											
授课学时分配	课堂讲授32学时											
先修课程	储能热流基础（含实验），储能化学基础（含实验），大学物理（含实验）											
后续课程	无											
教学方式	课堂讲授，文献阅读与小组讨论，大作业											
考核方式	课程结束笔试成绩占60%，大作业成绩占30%，平时成绩占10%											
参考教材	[1] 王艳艳，徐丽，李星国. 氢气储能与发电开发. 北京：化学工业出版社，2017. [2] 毛宗强，毛志明，余皓. 制氢工艺与技术. 北京：化学工业出版社，2018. [3] O'Hayre R, Cha S W, Colella W, et al. Fuel Cell Fundamentals. Hoboken: John Wiley & Sons,Inc., 2016.											
参考资料	[1] 蔡颖，许剑轶，胡锋，等. 储氢技术与材料. 北京：化学工业出版社，2018. [2] Dicks A L, Rand D A J. Fuel Cell Systems Explained. Hoboken: John Wiley & Sons, Inc., 2018.											

5.7.3 教学目的和基本要求（Teaching Objectives and Basic Requirements）

（1）熟悉氢经济、制氢技术的分类和发展趋势；

（2）掌握储氢的方法、优缺点和发展趋势；

（3）熟悉液氢的生产、储存和运输过程；

（4）掌握氢燃料电池的基本原理和技术、发展历史和应用现状。

（1）Be acquainted with the hydrogen economy, the hydrogen production technology, and development trends.

（2）Master the methods of hydrogen storage, including its advantages, and disadvantages as well as the corresponding development trends.

（3）Be acquainted with liquid hydrogen technology from production, storage to transportation.

（4）Master the hydrogen fuel cell principle and technology as well as its history and application status.

5.7.4 课程大纲和知识点（Syllabus and Key Points）

第一章 绪论（introduction）

章节序号 chapter number	章节名称 chapters	知识点 key points	课时 class hour
1.1	氢能概述 overview of hydrogen energy	（1）氢气的性质 （2）零排放的氢能经济 （3）氢能的制备、储存、运输和利用 （4）储氢的重要性 （5）氢政策与氢安全 （1）properties of hydrogen （2）zero-emission hydrogen energy economy （3）production, storage, transportation and utilization of hydrogen energy （4）importance of hydrogen storage （5）hydrogen policy and hydrogen safety	2
1.2	氢能的开发与利用 development and utilization of hydrogen energy	（1）氢能的开发现状 （2）直接燃烧技术 （3）燃料电池技术 （4）核聚变技术 （1）development of hydrogen energy （2）direct combustion technology （3）fuel cell technology （4）nuclear fusion technology	2

第二章　制氢技术简介（brief introduction of hydrogen production technology）

章节序号 chapter number	章节名称 chapters	知识点 key points	课时 class hour
2.1	制氢概述 overview of hydrogen production	（1）制氢方法的分类 （2）制氢技术的发展 （1）types of hydrogen production method （2）development of hydrogen production technology	0.5
2.2	化石燃料制氢 hydrogen production from fossil fuels	（1）煤制氢 （2）天然气制氢 （3）石油制氢 （1）hydrogen production from coal （2）hydrogen production from natural gas （3）hydrogen production from petroleum	1
2.3	可再生能源制氢 hydrogen production from renewable energy	（1）太阳能制氢 （2）生物质制氢 （3）风能制氢 （1）hydrogen production from solar energy （2）hydrogen production from biomass （3）hydrogen production from wind energy	1
2.4	其他制氢方式 other hydrogen production methods	（1）甲醇制氢 （2）甲烷制氢 （3）氨气制氢 （4）金属粉末制氢 （5）副产氢的回收与纯化 （1）hydrogen production from methanol （2）hydrogen production from methane （3）hydrogen production from ammonia （4）hydrogen production from metal powder （5）recovery and purification of byproduct hydrogen	1.5

第三章　储氢技术（hydrogen storage technology）

章节序号 chapter number	章节名称 chapters	知识点 key points	课时 class hour
3.1	储氢概述 overview of hydrogen storage	（1）气态储氢 （2）液态储氢 （3）固态储氢 （1）gaseous hydrogen storage （2）liquid hydrogen storage （3）solid hydrogen storage	2

续表

章节序号 chapter number	章节名称 chapters	知识点 key points	课时 class hour
3.2	储氢材料 hydrogen storage materials	（1）储氢材料的定义 （2）储氢材料的要求 （3）储氢材料的类型 （4）储氢材料的性能及经济性分析 （1）definition of hydrogen storage materials （2）requirements for hydrogen storage materials （3）types of hydrogen storage materials （4）performance and economic analysis of hydrogen storage materials	3
3.3	储氢容器 hydrogen storage vessels	（1）储氢罐的分类 （2）储氢罐的发展状况 （3）储氢容器经济性分析 （1）types of hydrogen storage vessel （2）development of hydrogen storage vessel （3）economic analysis of hydrogen storage vessel	3

第四章　液氢技术（liquid hydrogen technology）

章节序号 chapter number	章节名称 chapters	知识点 key points	课时 class hour
4.1	液氢的背景与性质 background and properties of liquid hydrogen	（1）液氢的物性 （2）液氢的外延产品 （3）液氢的用途 （4）液氢的安全 （1）physical properties of liquid hydrogen （2）extended products of liquid hydrogen （3）applications of liquid hydrogen （4）safety of liquid hydrogen	1
4.2	液氢的生产 liquid hydrogen production	（1）液氢的生产工艺与流程 （2）液氢的生产成本 （1）production process of liquid hydrogen （2）production cost of liquid hydrogen	2
4.3	液氢的储存与运输 storage and transportation of liquid hydrogen	（1）液氢储罐运输 （2）液氢管道输送 （3）液氢加注系统 （1）storage tank transportation of liquid hydrogen （2）pipeline transportation of liquid hydrogen （3）filling system of liquid hydrogen	1

第五章 氢燃料电池原理（principle of hydrogen fuel cell）

章节序号 chapter number	章节名称 chapters	知识点 key points	课时 class hour
5.1	燃料电池简介 overview of fuel cell	（1）燃料电池 （2）燃料电池的优缺点 （3）燃料电池的类型 （4）燃料电池的基本工作原理 （5）燃料电池性能 （6）燃料电池的应用 （1）fuel cell （2）advantages and disadvantages of fuel cell （3）types of fuel cell （4）principle of fuel cell （5）performance of fuel cell （6）applications of fuel cell	2
5.2	热力学与动力学 thermodynamics and kinetics	（1）反应焓 （2）吉布斯自由能 （3）活化能 （4）能斯特方程 （5）交换电流密度 （6）巴特勒-福尔默方程 （7）塔菲尔等式 （8）电催化剂 （1）reaction enthalpy （2）Gibbs free energy （3）activation energy （4）Nernst equation （5）exchange current density （6）Butler-Volmer Equation （7）Tafel equation （8）electrocatalyst	2
5.3	传质与传荷 mass and charge transport	（1）电荷迁移 （2）电解质 （3）流场中的对流传输 （4）电极中的扩算传输 （1）charge mobility （2）electrolyte （3）convective mass transport in flow field （4）diffusive transport in electrode	2

第六章　氢燃料电池技术（hydrogen fuel cell technology）

章节序号 chapter number	章节名称 chapters	知识点 key points	课时 class hour
6.1	膜电极 membrane electrode assembly	（1）膜电极的结构与组成 （2）膜电极的性能 （3）膜电极的表征 （1）structure and component of membrane electrode assembly （2）performance of membrane electrode assembly （3）characterization of membrane electrode assembly	2
6.2	燃料电池电堆 fuel cell stack	（1）电堆的结构与组成 （2）水热管理 （3）电堆的功率 （4）电堆的效率 （1）structure and component of stack （2）thermal and water management （3）stack power （4）stack efficiency	2
6.3	系统集成及应用 system integration and application	（1）燃料电池发动机 （2）燃料电池汽车 （3）燃料电池电站 （1）fuel cell engine （2）fuel cell vehicle （3）fuel cell power plant	2

第 6 章

电化学与电磁储能课程群

6.1 "储能材料工程"教学大纲

课程名称：储能材料工程

Course Title：Energy Storage Material Engineering

先修课程：电化学基础，储能原理 I

Prerequisites：Fundamentals of Electrochemistry，Energy Storage Principle I

学分：3

Credits：3

6.1.1 课程目的和基本内容（Course Objectives and Basic Content）

本课程是储能科学与工程专业的一门必修课。

本课程从物质结构的基础出发，讲述储能材料从原子（或离子、分子）、原子间作用（结合方式）、原子间排列到储能材料基本特性和储能形式的基本概念和基本理论。并结合储能材料的表征及应用，使学生通过学习基础知识，深刻准确地理解基本核心概念，牢固掌握基本理论，从而培养学生利用理论知识解决实际工程问题的能力。

课程教学中，不仅讲授材料物质结构的基础知识、储能材料的基本特性、储能原理，而且强调高分子储能材料的特性、热物性及测定方法和遴选原则，使学生在掌握储能材料基本概念的同时，理解储能材料性质与性能之间的构效关系，为储能材料的设计、制备、改性及应用奠定基础。通过文献调研训练，使学生了解储能材料的最新研究动态，建立终身学习的意识，并掌握查阅文献资料的能力，不断追求最新的知识理论，发展和完善知识体系。结合储能体系的发展前沿，培养学生发现问题、解决问题的能力。

This course is a mandatory course for undergraduates majoring in energy storage science and engineering.

This course mainly consists of three parts. The first part is introduction of basic concepts and theories of material structure, including atoms (or ions, molecules), the interaction between atoms (combination method), and arrangement of atoms. The second part introduces the forms and concepts of energy storage materials. The third part focuses on basic characteristics of energy storage materials. In addition, combined with the characterization and application of energy storage materials, this course helps students systematically understand the basic theory of energy storage materials and cultivate the ability to solve practical engineering problems.

In this course, all the basic ideas and common knowledge of structure, characteristics, and properties of energy storage materials are taught. Students can master the basics of energy storage materials. Additionally, the course enables students to understand the structure-activity relationship between the properties and performance of energy storage materials to lay a good foundation for the design, preparation, modification and application of energy storage materials. Through literature, students can understand the latest research trends of energy storage materials, establish a sense of lifelong learning, master the ability to consult the literature, constantly pursue the latest knowledge and theories, and develop their knowledge system. Therefore, mastering the basic concept, basic principles, and methods of this course will make it easier for students to solve practical problems of energy storage system in the future.

6.1.2　课程基本情况（Basic Information of the Course）

课程名称	储能材料工程 Energy Storage Material Engineering										
开课时间	一年级		二年级		三年级		四年级		总学分	3	总学时	56
	秋	春	秋	春	秋	春	秋	春				
课程定位	储能科学与工程专业本科生电化学与电磁储能课程群（B模块）必修课											
授课学时分配	课堂讲授40学时+实验16学时											
先修课程	电化学基础,储能原理Ⅰ											
后续课程	热质储能技术及应用（含实验）,能源互联网,电池材料制备技术,新型储能电池											
教学方式	课堂教学,大作业,实验与综述报告,平时作业											
考核方式	课程结束笔试成绩占60%,实验报告占20%,平时成绩占20%											
参考教材	[1] 吴贤文,向延鸿. 储能材料:基础与应用. 北京:化学工业出版社,2019.											
参考资料	[1] 樊栓狮,梁德青,杨向阳. 储能材料与技术. 北京:化学工业出版社,2004. [2] 布鲁内特.储能技术与应用. 唐国胜,译. 北京:机械工业出版社,2018. [3] 云斯宁. 新型能源材料与器件. 北京:中国建材工业出版社,2019.											

6.1.3　教学目的和基本要求（Teaching Objectives and Basic Requirements）

（1）具备储能材料基础知识的理解能力。掌握储能材料的基础知识,包括物质结构、储能材料的基本特性、储能形式;掌握储能材料物理、化学性质,以及各个性质之间的相互关系和影响因素,并掌握材料结构、性质与性能之间的构效关系。

（2）具备储能材料工程问题分析能力。能够运用储能材料的相关知识，对储能材料在实际应用中的具体工程问题进行分析。能够对比分析不同体系、不同类型的储能材料或制备工艺获得的储能材料的成分、结构、性质的特点，提出有效的解决方案和途径。

（3）具备利用专业知识结合文献研究解决复杂工程问题的能力。针对储能材料领域的复杂工程问题，能够利用储能材料工程的基本知识，通过文献检索、调研等科学方法进行研究，合理选择材料体系，制订研究路线，设计实验方案，安全有效地开展实验，最终获得合理有效的结论或解决办法。

（1）Ability to understand the basic knowledge of energy storage materials. In this course, students should master the basic knowledge of energy storage materials, including material structure, basic characteristics of energy storage materials, and energy storage forms. In addition, students should master the physical and chemical properties of energy storage materials, as well as the interrelationships and influencing factors between various properties, and understand the structure, properties, and structure-activity relationship of materials.

（2）Ability to analyze the engineering problems of energy storage materials. Students can use the relevant knowledge of energy storage materials to analyze specific engineering problems in practical applications. Furthermore, students can compare and analyze the characteristics of the composition, structure, and properties of energy storage materials obtained by various systems and types of energy storage materials or preparation processes. And they can use what they learn to propose effective approaches and solutions.

（3）Ability to use basic knowledge and literature to solve complex engineering problems.Facing the complex engineering problems in energy storage materials, students can use the basic knowledge of energy storage materials to conduct research. Furthermore, combined with literature retrieval and research, they can rationally select material systems, formulate research routes, design experimental plans, conduct experiments, and finally resolve the problems.

6.1.4　课程大纲和知识点（Syllabus and Key Points）

第一章　绪论（introduction）

章节序号 chapter number	章节名称 chapters	知识点 key points	课时 class hour
1.1	能源与储能材料 energy and energy storage materials	（1）能源概述 （2）储能材料概述 （1）introduction of energy （2）introduction of energy materials	1

续表

章节序号 chapter number	章节名称 chapters	知识点 key points	课时 class hour
1.2	储能材料的主要应用现状与进展 application and progress of energy storage materials	（1）储能材料研究现状 （2）储能材料发展趋势 （1）research status of energy storage materials （2）trend of energy storage materials	1

第二章　物质结构基础（fundamentals of material structure）

章节序号 chapter number	章节名称 chapters	知识点 key points	课时 class hour
2.1	原子结构与性质 atomic structure and properties	（1）原子结构 （2）原子结合键 （3）原子排布方式 （1）atomic structure （2）atomic bond （3）atomic arrangement	1
2.2	晶体学基础 fundamentals of crystallography	（1）晶系与点阵 （2）晶向指数与晶面指数 （3）晶带及晶带定理 （4）纯金属的晶体结构 （5）离子晶体的结构 （6）共价晶体的结构 （1）crystal system and lattice （2）crystal orientation index and lattice plane index （3）crystal ribbon and crystal ribbon theorem （4）crystal structure of pure metal （5）structure of ionic crystals （6）structure of covalent crystals	2
2.3	晶体形貌与生长 morphology and growth of crystal	（1）晶体生长的动力学原理 （2）晶体生长过程的物相及热力学描述 （3）晶体生长过程的形核原理 （1）kinetic principle of crystal growth （2）phase and thermodynamic description of crystal growth process （3）nucleation principle of crystal growth process	2
2.4	晶体缺陷 crystal defects	（1）点缺陷 （2）位错的基本概念 （3）位错的能量和交互作用	2

续表

章节序号 chapter number	章节名称 chapters	知识点 key points	课时 class hour
2.4	晶体缺陷 crystal defects	（4）晶体中的界面 （1）point defects （2）basic concept of dislocation （3）energy and interaction of dislocations （4）interface in the crystal	2
2.5	X衍射定律和衍射几何 X diffraction law and diffraction geometry	（1）X射线特征谱 （2）劳埃方程和布拉格方程 （3）衍射消光和衍射强度 （1）X-ray characteristic spectrum （2）Laue's equation and Bragg's equation （3）diffraction extinction and diffraction intensity	1

第三章 储能的形式及材料（forms of energy storage and energy storage materials）

章节序号 chapter number	章节名称 chapters	知识点 key points	课时 class hour
3.1	化学储能 chemical energy storage	（1）化学储能的概念及特征 （2）化学储能的主要指标 （3）化学储能的主要用途 （4）化学储能材料 （1）concept and characteristics of chemical energy storage （2）main metrics of chemical energy storage （3）main applications of chemical energy storage （4）chemical energy storage materials	4
3.2	机械储能 mechanical energy storage	（1）机械储能的概念及特征 （2）机械储能的主要指标 （3）机械储能的主要用途 （4）机械储能材料 （1）concept and characteristics of mechanical energy storage （2）main metrics of mechanical energy storage （3）main applications of mechanical energy storage （4）mechanical energy storage materials	3
3.3	电磁储能 electromagnetic energy storage	（1）电磁储能的概念及特征 （2）电磁储能的主要指标 （3）电磁储能的主要用途 （4）电磁储能材料 （1）concept and characteristics of electromagnetic energy storage	2.5

续表

章节序号 chapter number	章节名称 chapters	知识点 key points	课时 class hour
3.3	电磁储能 electromagnetic energy storage	（2）main metrics of electromagnetic energy storage （3）main applications of electromagnetic energy storage （4）electromagnetic energy storage materials	2.5
3.4	热能存储 thermal energy storage	（1）热能存储的概念及特征 （2）热能存储的主要指标 （3）热能存储的主要用途 （4）热能存储材料 （1）concept and characteristics of thermal energy storage （2）main metrics of thermal energy storage （3）main applications of thermal energy storage （4）thermal energy storage materials	2
3.5	储能技术的特点和总结 characteristics and summary of energy storage technology	（1）储能技术的特点 （2）储能的重大战略意义 （1）characteristics of energy storage technology （2）strategic significance of energy storage	0.5

第四章 储能材料的基本特性（characteristics of energy storage materials）

章节序号 chapter number	章节名称 chapters	知识点 key points	课时 class hour
4.1	相变的焓差 enthalpy difference of phase transition	（1）相变焓的定义及公式 （2）相变的影响因素 （1）definition and formula of phase transition enthalpy （2）factors to influence phase transition	2
4.2	相平衡特性 phase equiliubrium properties	（1）相图及相律 （2）熵增原理 （3）相平衡条件 （1）phase diagram and phase rule （2）principle of entropy increase （3）phase equilibrium conditions	3
4.3	相变过程特性 properties of phase transition process	（1）结晶过程 （2）晶体生长的影响因素 （3）成核材料简介 （1）crystallization process （2）factors to influence crystal growth （3）introduction to nucleation materials	2

续表

章节序号 chapter number	章节名称 chapters	知识点 key points	课时 class hour
4.4	高分子储能材料特性 properties of polymer energy storage materials	（1）高分子相变储能材料特性 （2）高分子储能材料举例 （1）properties of polymer phase change energy storage materials （2）examples of polymer energy storage materials	1
4.5	储能材料的热物性及测定方法 thermal properties and measurement methods of energy storage materials	（1）储能材料的热物性定义 （2）相变温度和相变潜热的测定方法 （1）definition of thermal properties of energy storage materials （2）measurement methods of phase transition temperature and latent heat of phase transition	1
4.6	储能材料的遴选原则 selection principles for energy storage materials	（1）储能材料的分类 （2）常用材料储能特性对比 （1）classification of energy storage materials （2）comparison of energy storage characteristics for commonly used materials	1

第五章 储能材料的应用（application of energy storage materials）

章节序号 chapter number	章节名称 chapters	知识点 key points	课时 class hour
5.1	一次电池 primary batteries	（1）一次电池的工作原理 （2）一次电池的特点 （3）一次电池的种类 （1）working principle of primary batteries （2）characteristics of primary batteries （3）types of primary batteries	1
5.2	二次电池 secondary batteries	（1）二次电池的工作原理 （2）二次电池的特点 （3）二次电池评价指标 （4）锂离子电池常用材料 （1）working principle of secondary batteries （2）characteristics of secondary batteries （3）evaluation index of secondary batteries （4）commonly used materials for lithium-ion batteries	2

续表

章节序号 chapter number	章节名称 chapters	知识点 key points	课时 class hour
5.3	燃料电池 fuel cells	（1）燃料电池的工作原理 （2）燃料电池的特点 （3）碱性/磷酸盐燃料电池材料 （1）working principle of fuel cells （2）characteristics of fuel cells （3）alkaline/phosphate fuel cells	2
5.4	太阳能电池 solar cells	（1）太阳能电池的工作原理 （2）太阳能电池的特点 （3）半导体材料 （1）working principle of solar cells （2）characteristics of solar cells （3）semiconductors	1.5
5.5	液流电池 flow batteries	（1）液流电池工作原理 （2）液流电池评价指标 （3）液流电池材料要求 （1）working principle of flow batteries （2）evaluation index of flow batteries （3）requirements of flow battery materials	1
5.6	储氢材料 hydrogen storage materials	（1）储氢材料的基本概念 （2）储氢材料简介 （1）basic concepts of hydrogen storage materials （2）introduction of hydrogen storage materials	0.5

6.1.5　实验环节（Experiments）

序号 number	实验内容 experiment content	知识点 key points	课时 class hour
1	超级电容器 supercapacitor	（1）组装超级电容器 （2）性能测试 （3）电化学数据分析 （1）supercapacitor assembly （2）test of the electrochemical performance （3）electrochemical data analysis	5

续表

序号 number	实验内容 experiment content	知识点 key points	课时 class hour
2	锂离子电池 lithium-ion battery	（1）组装扣式电池 （2）电池测试 （3）电化学数据分析 （1）coin cell assembly （2）test of the electrochemical performance （3）electrochemical data analysis	6
3	液流电池 flow battery	（1）搭建装置 （2）电池测试 （3）电化学数据分析 （1）device setup （2）test of the electrochemical performance （3）electrochemical data analysis	5

6.2 "电化学基础"教学大纲

课程名称：电化学基础

Course Title：Fundamentals of Electrochemistry

先修课程：储能化学基础（含实验）

Prerequisites：General Chemistry for Energy Storage（with Experiments）

学分：3

Credits：3

6.2.1 课程目的和基本内容（Course Objectives and Basic Contents）

本课程是储能科学与工程专业的一门专业核心课。

电化学基础的理论在纳米材料制备、锂离子电池、钠离子电池、锂硫电池、锂空电池、钠空电池、超级电容器、电解电容器、液流电池、太阳能电池、光催化、电催化、氢气制备、燃料电池、石墨烯等二维材料制备、金属腐蚀和防腐等重要领域中具有广泛的应用。随着电化学研究在多学科交叉研究中日益受到关注，课程将从电化学的实际应用出发，介绍具有新工科特色的电化学基础原理及应用知识，从而更好地为各学科交叉研究提供系统的电化学知识。

This course is a professional core course for undergraduates majoring in energy storage science and engineering.

The fundamentals of electrochemistry have a wide range of applications in several important fields such as the preparation of nanomaterials, lithium-ion batteries, sodium-ion batteries, lithium-sulfur batteries, lithium-air batteries, sodium-air batteries, supercapacitors, electrolytic capacitors, flow batteries, solar cells, photocatalysis, electrocatalysis, hydrogen preparation, fuel cells, preparation of graphene and other two-dimensional materials, metal corrosion and anti-corrosion. As electrochemical research has received increasing attention in interdisciplinary research, the course will start from the practical application of electrochemistry, introduce the basic principles and application knowledge of electrochemistry with new engineering characteristics, to better facilitate interdisciplinary research via knowledge of electrochemistry.

6.2.2　课程基本情况（Basic Information of the Course）

课程名称	电化学基础 Fundamentals of Electrochemistry											
开课时间	一年级		二年级		三年级		四年级		总学分	3	总学时	56
	秋	春	秋	春	秋	春	秋	春				
课程定位	储能科学与工程专业本科生电化学与电磁储能课程群必修课											
授课学时分配	课堂讲授40学时+实验16学时											
先修课程	储能化学基础（含实验）											
后续课程	储能材料工程,电池材料制备技术,新型储能电池技术,固态电池,储能电池设计、制作及集成化实验											
教学方式	课堂教学,大作业,实验与综述报告,平时作业											
考核方式	课程结束笔试成绩占70%,实验报告占20%,平时成绩占10%											
参考教材	[1] 杨文治. 电化学基础. 北京:北京大学出版社,1983.											
参考资料	[1] Bard A J, Faulkner L R. Electrochemical Methods Fundamentals and Applications. Hoboken: John Wiley & Sons, Inc., 2001. [2] Hamann C H .电化学. 2版. 陈艳霞,等,译. 北京:化学工业出版社,2010.											

6.2.3　教学目的和基本要求（Teaching Objectives and Basic Requirements）

通过本课程的学习，使学生能够从电化学的实际应用出发，了解且掌握具有新工科特色的电化学基础原理及应用知识，更好地为各学科交叉研究提供系统的电化学知识。

基本要求：

（1）了解电的起源、发展以及电化学发展及应用历史；

（2）通过锂离子电池的应用实例，了解电化学中的基本概念和术语；

（3）了解液体、胶体和固体三大类电解质体系，掌握电解质的基本原理和特性；

（4）掌握电化学过程中的热力学基本原理；

（5）掌握电化学过程中的动力学基本原理；

（6）掌握当前常用的电化学研究方法；

（7）了解电化学制备和加工的典型应用；

（8）了解电化学器件的典型应用及其特性；

（9）了解电化学在其他方面的前沿应用。

Through the study of this course, students can start from the practical application of electrochemistry, understand and master the basic principles and application knowledge of electrochemistry with new engineering characteristics, and master systematic electrochemical knowledge for interdisciplinary research.

Basic requirements:

（1）Understand the origin and development of electricity and the development and application history of electrochemistry.

（2）Though the practical application examples of lithium-ion batteries, understand the basic concepts and terms in electrochemistry.

（3）Understand the three major electrolyte systems of liquid, colloid, and solid, and understand the basic principles and characteristics of electrolytes.

（4）Master the basic principles of thermodynamics in electrochemical processes.

（5）Master the basic principles of kinetics in electrochemical processes.

（6）Master the current commonly used electrochemical research methods.

（7）Understand the typical applications of electrochemical preparation and processing.

（8）Understand the typical applications and characteristics of electrochemical devices.

（9）Understand the cutting-edge applications of electrochemistry in other areas.

6.2.4 课程大纲和知识点（Syllabus and Key Points）

第一章 绪论（introduction）

章节序号 chapter number	章节名称 chapters	知识点 key points	课时 class hour
1.1	电化学问题 electrochemical problems	（1）电的起源、发展 （2）电化学发展及应用历史 （1）origin and development of electricity （2）development and application history of electrochemistry	1

章节序号 chapter number	章节名称 chapters	知识点 key points	课时 class hour
1.2	其他电化学过程 other electrochemical processes	其他电化学过程 other electrochemical processes	1

第二章　电化学基本概念的引出（introduction of the basic concepts of electrochemistry）

章节序号 chapter number	章节名称 chapters	知识点 key points	课时 class hour
2.1	锂离子电池的实际应用 practical application of lithium-ion battery	（1）锂离子电池的发展 （2）锂离子电池概况 （3）锂离子电池正极材料 （4）负极材料 （5）电解质 （6）锂离子电池的制造过程 （7）锂离子电池的应用 （1）development of lithium-ion batteries （2）overview of lithium-ion batteries （3）cathode material of lithium-ion batteries （4）anode materials （5）electrolyte （6）manufacturing process of lithium-ion battery （7）application of lithium-ion battery	4
2.2	电化学中的基本概念和术语 basic concepts and terms in electrochemistry	电化学中的基本概念和术语 basic concepts and terms in electrochemistry	1

第三章　电解质理论（electrolyte theory）

章节序号 chapter number	章节名称 chapters	知识点 key points	课时 class hour
3.1	液体、胶体和固体三大类电解质体系 three types of electrolyte systems: liquid, colloid, and solid	液体、胶体和固体三大类电解质体系 three types of electrolyte systems: liquid, colloid, and solid	1

续表

章节序号 chapter number	章节名称 chapters	知识点 key points	课时 class hour
3.2	电解质的基本原理和特性 basic principles and characteristics of electrolytes	（1）溶液导电的本质 （2）电导法的应用 （3）电解质的平衡性质 （1）nature of solution conductivity （2）application of conductance technique （3）equilibrium properties of electrolyte	2

第四章　电化学热力学（electrochemical thermodynamics）

章节序号 chapter number	章节名称 chapters	知识点 key points	课时 class hour
4.1	电化学过程中的热力学基本原理 basic principles of thermodynamics in electrochemical processes	电化学过程中的热力学基本原理 basic principles of thermodynamics in electrochemical processes	
4.2	化学能与电能的相互转化 interconversion between chemical energy and electric energy	（1）化学能 （2）化学能与电能的相互转化 （3）化学电源 （4）电解电池 （1）chemical energy （2）interconversion between chemical energy and electric energy （3）chemical power （4）electrolytic cell	1
4.3	电池的平衡电势 equilibrium potential of the battery	（1）可逆电池 （2）可逆电极 （3）金属电极 （4）氧化还原电极 （5）气体电极 （1）reversible battery （2）reversible electrode （3）metal electrode （4）redox electrode （5）gas electrode	1

章节序号 chapter number	章节名称 chapters	知识点 key points	课时 class hour
4.4	Nernst公式 Nernst formula	（1）电池电动势 （2）电池电动势与溶液浓度和温度、压力等的关系 （1）battery electromotive force （2）relationship between battery electromotive force and solution concentration, temperature, pressure, etc.	1
4.5	电池电动势与温度的关系 relationship between battery electromotive force and temperature	电池电动势与温度的关系 relationship between battery electromotive force and temperature	
4.6	电势法的应用 application of potential method	（1）电池电动势的测量 （2）电势法 （1）measurement of battery electromotive force （2）potential method	1
4.7	电极的平衡电势 equilibrium potential of electrode	（1）电极电势 （2）标准氢电极 （3）标准电极电势表 （4）电极电势和浓度的关系 （1）electrode potential （2）standard hydrogen electrode （3）standard electrode potential table （4）relationship between electrode potential and concentration	1
4.8	有液体接界面的电池 batteries with liquid contact interface	（1）扩散电势 （2）液体接界电势 （3）盐桥 （1）diffusion potential （2）liquid junction potential （3）salt bridge	1
4.9	电势法测量电解质 potentiometric measurement of electrolyte	电势法测量电解质 potentiometric measurement of electrolyte	
4.10	pH的意义和pH的电势测量法 meaning of pH and the potential measurement method of pH	pH的意义和pH的电势测量法 meaning of pH and the potential measurement method of pH	1

第五章 电化学动力学（electrochemical kinetics）

章节序号 chapter number	章节名称 chapters	知识点 key points	课时 class hour
5.1	电化学过程中的动力学基本原理 basic principles of kinetics in electrochemical processes	电化学过程中的动力学基本原理 basic principles of kinetics in electrochemical processes	1
5.2	双电层结构 electric double-layer structure	（1）离子双电层 （2）表面张力 （3）电荷密度和电容 （4）双电层模型 （1）ionic electric double layer （2）surface tension （3）charge density and capacitance （4）electric double-layer model	1
5.3	巴特勒-福尔默方程及其应用 Butler-Volmer equation and its application	（1）巴特勒-福尔默方程 （2）巴特勒-福尔默方程的应用 （1）Butler-Volmer equation （2）application of Butler-Volmer equation	1

第六章 电化学研究方法（electrochemical research methods）

章节序号 chapter number	章节名称 chapters	知识点 key points	课时 class hour
6.1	当前常用的电化学研究方法 current commonly used electrochemical research methods	当前常用的电化学研究方法 current commonly used electrochemical research methods	1
6.2	电化学问题的数值计算方法 numerical calculation methods for electrochemical problems	电化学问题的数值计算方法 numerical calculation methods for electrochemical problems	1
6.3	稳态研究方法 steady-state research methods	稳态研究方法 steady-state research methods	1
6.4	暂态研究方法 transient research methods	暂态研究方法 transient research methods	1

章节序号 chapter number	章节名称 chapters	知识点 key points	课时 class hour
6.5	电化学测试方法和仪器 electrochemical test methods and instruments	电化学测试方法和仪器 electrochemical test methods and instruments	1

第七章　电化学制备与加工（electrochemical preparation and processing）

章节序号 chapter number	章节名称 chapters	知识点 key points	课时 class hour
7.1	电化学制备和加工的典型应用 typical applications of electrochemical preparation and processing	电化学制备和加工的典型应用 typical applications of electrochemical preparation and processing	1
7.2	纳米材料制备 preparation of nanomaterials	纳米材料制备 preparation of nanomaterials	1
7.3	石墨烯等二维材料制备 preparation of two-dimensional materials such as graphene	石墨烯等二维材料制备 preparation of two-dimensional materials such as graphene	1
7.4	电化学聚合 electrochemical polymerization	电化学聚合 electrochemical polymerization	1
7.5	电化学金属表面处理 surface treatment of electrochemical metal	电化学金属表面处理 surface treatment of electrochemical metal	1
7.6	金属腐蚀和防腐、电镀、电解加工等 metal corrosion and anti-corrosion, electroplating, electrolytic processing	金属腐蚀和防腐、电镀、电解加工等 metal corrosion and anti-corrosion, electroplating, electrolytic processing	1

第八章 电化学器件（electrochemical device）

章节序号 chapter number	章节名称 chapters	知识点 key points	课时 class hour
8.1	电化学器件的典型应用及其特性 typical applications and characteristics of electrochemical devices	电化学器件的典型应用及其特性 typical applications and characteristics of electrochemical devices	1
8.2	锂/钠离子电池 lithium/sodium ion batteries	锂/钠离子电池 lithium/sodium ion batteries	1
8.3	超级电容器 supercapacitors	超级电容器 supercapacitors	
8.4	电解电容器 electrolytic capacitors	电解电容器 electrolytic capacitors	1
8.5	液流电池 flow batteries	液流电池 flow batteries	
8.6	太阳能电池 solar batteries	太阳能电池 solar batteries	
8.7	燃料电池 fuel cells	燃料电池 fuel cells	1
8.8	电化学传感器 electrochemical sensors	电化学传感器 electrochemical sensors	

第九章 电化学的其他应用（other applications of electrochemistry）

章节序号 chapter number	章节名称 chapters	知识点 key points	课时 class hour
9.1	电化学在其他方面的前沿应用 frontier applications of electrochemistry in other areas	电化学在其他方面的前沿应用 frontier applications of electrochemistry in other areas	1
9.2	光催化 photocatalytic	光催化 photocatalytic	1
9.3	电催化 electrocatalysis	电催化 electrocatalysis	1

章节序号 chapter number	章节名称 chapters	知识点 key points	课时 class hour
9.4	光解水制氢 hydrogen production by photolysis of water	光解水制氢 hydrogen production by photolysis of water	1
9.5	电去离子水处理 electrodeionized water treatment	电去离子水处理 electrodeionized water treatment	1
9.6	电化学检测 electrochemical detection	电化学检测 electrochemical detection	
9.7	氧泵 oxygen pump	氧泵 oxygen pump	

6.2.5　实验环节（Experiments）

序号 number	实验内容 experiment content	知识点 key points	课时 class hour
1	电解液电导率测试 electrolyte conductivity test	（1）电解液电导率的概念和特性 （2）电导仪及电化学工作站的使用 （1）concept and characteristics of electrolyte conductivity （2）use of conductivity meter and electrochemical workstation	4
2	电池材料的循环伏安测试 cyclic voltammetry test of battery materials	（1）电池材料的定义、结构及充放电原理 （2）电池材料的充放电过程及特性 （3）电化学工作站的使用 （1）definition, structure, and principle of charge and discharge of battery materials （2）charging and discharging process and characteristics of battery materials （3）use of electrochemical workstation	4
3	电池材料的电化学阻抗谱测试 electrochemical impedance spectroscopy test of battery materials	（1）电池材料电化学阻抗谱的定义及测量原理 （2）电化学工作站的使用 （1）definition and measurement principle of electrochemical impedance spectroscopy of battery materials （2）use of electrochemical workstation	4

续表

序号 number	实验内容 experiment content	知识点 key points	课时 class hour
4	电池材料比容量和库仑效率测试 specific capacity and Coulomb efficiency test of battery materials	电池材料比容量和库仑效率的定义及测试原理 definition and measurement principle of specific capacity and Coulomb efficiency of battery materials	4

6.3 "半导体物理"教学大纲

课程名称：半导体物理
Course Title：Semiconductor Physics
先修课程：大学物理（含实验）
Prerequisites：Physics and Physics Experiments
学分：3
Credits：3

6.3.1 课程目的和基本内容（Course Objectives and Basic Contents）

本课程是储能科学与工程专业的一门专业核心课。

通过学习，要求学生全面了解和掌握半导体物理的基本知识和基础理论，半导体器件的特性、工作原理及其局限性，为后继专业课程的学习以及将来从事储能科学与工程研究和半导体及集成电路的工艺和设计等工作奠定基础。

主要内容有：（1）基础物理知识，包括固体晶体结构、量子力学和固体物理；半导体材料物理知识，主要是平衡态和非平衡态半导体以及载流子输运现象，包括半导体中的电子状态、半导体中的杂质和缺陷能级、载流子的统计分布、半导体的导电性、非平衡载流子。（2）半导体器件物理知识，包括同质 PN 结、金属半导体接触、异质结、双极晶体管、MOS 场效应晶体管、结型场效应晶体管、半导体光电器件以及功率半导体器件等。

This course is a professional core course for undergraduates majoring in energy storage science and engineering.

Through learning, students are required to fully understand and master the basic knowledge and basic theory of semiconductor physics, the characteristics, working principles, and limitations of semiconductor devices, for the subsequent professional courses.This course lays the foundation for research in the field of energy storage science and engineering, and the process and design of semiconductor and integration circuit in the future.

The main content is basic physics knowledge, including solid crystal structure, quantum mechanics and solid physics. The course also covers physics knowledge of semiconductor materials, including equilibrium and non-equilibrium semiconductors and carrier transport phenomena, which include electronic states in semiconductors, impurities and defect energy levels in semiconductors, the statistical distribution of carriers, conductivity of semiconductors, and non-equilibrium carriers. Last but not the least, it introduces the physics of semiconductor device, including homogenous PN junction, metal-semiconductor contact, heterojunction, bipolar transistor, MOS field-effect transistor, junction field-effect transistors, semiconductor optoelectronic devices, power semiconductor devices, etc.

6.3.2　课程基本情况（Basic Information of the Course）

课程名称	半导体物理 Semiconductor Physics										
开课时间	一年级		二年级		三年级		四年级		总学分	3	总学时	56
	秋	春	秋	春	秋	春	秋	春			
课程定位	储能科学与工程专业本科生电化学与电磁储能课程群必修课										
授课学时分配	课堂讲授40学时+实验16学时										
先修课程	大学物理（含实验）										
后续课程	电池材料制备技术，纳米材料与能源，新型储能电池技术，固态电池，储能电池设计、制作及集成化实验										
教学方式	课堂教学，实验与平时作业										
考核方式	课程结束笔试成绩占70%，期中考试占5%，平时成绩占25%										
参考教材	[1] Neamen D A.Semiconductor Physics and Devices: Basic Principle.New York：McGraw-Hill Companies,Inc.,2003. [2] Neamen D A. 半导体物理与器件. 4版.赵毅强，姚素英，史再峰，译.北京: 电子工业出版社,2018.										
参考资料	[1] 刘恩科，朱秉升，罗晋生. 半导体物理学.北京：电子工业出版社,2003.										

6.3.3　教学目的和基本要求（Teaching Objectives and Basic Requirements）

通过学习，要求学生全面地了解和掌握半导体物理的基本知识和基础理论，半导体器件的特性、工作原理及其局限性，为后继专业课程的学习，以及将来从事储能科学与工程研究和半导体及集成电路的工艺和设计等工作奠定基础。

基本要求：

（1）了解基础物理知识，包括固体晶体结构、量子力学和固体物理；

（2）了解半导体材料物理知识，主要是平衡态和非平衡态半导体以及载流子输运现象，包括半导体中的电子状态、半导体中的杂质和缺陷能级、载流子的统计分布、半导体的导电性、非平衡载流子；

（3）了解半导体器件物理知识，包括同质 PN 结、金属半导体接触、异质结、双极晶体管、MOS 场效应晶体管、结型场效应晶体管、半导体光电器件以及功率半导体器件等。

Through learning, students are required to fully understand and master the basic knowledge and basic theory of semiconductor physics, the characteristics, working principles, and limitations of semiconductor devices, for the subsequent professional courses. This course lays the foundation for research in the field of energy storage science and engineering, and the process and design of semiconductor and integration circuit in the future.

Basic requirements:

（1）Understand the basic physics knowledge, including solid crystal structure, quantum mechanics, and solid physics.

（2）Understand the physics of semiconductor materials, including equilibrium and non-equilibrium semiconductors and carrier transport phenomena. It includes electronic states in semiconductors, impurity and defect levels in semiconductors, statistical distribution of carriers, conductivity of semiconductors, and non-equilibrium carriers.

（3）Understand the physics of semiconductor device, including homogenous PN junction, metal-semiconductor contact, heterojunction, bipolar transistor, MOS field-effect transistor, junction field-effect transistor, semiconductor optoelectronic devices, and power semiconductor device etc.

6.3.4 课程大纲和知识点（Syllabus and Key Points）

绪论（introduction）

章节序号 chapter number	章节名称 chapters	知识点 key points	课时 class hour
0.1	半导体 semiconductors	半导体和集成电路的历史、器件 history and device of semiconductors and integrated circuits	0.5
0.2	集成电路 integrated circuits	集成电路(IC)和制造 integrated circuits (IC) and manufacturing	0.5

第一章　固体晶格结构（the crystal structure of solids）

章节序号 chapter number	章节名称 chapters	知识点 key points	课时 class hour
1.1	半导体材料 semiconductor materials	半导体材料 semiconductor materials	0.5
1.2	固体类型 types of solids	固体类型 types of solids	
1.3	空间晶格 space lattices	（1）原胞和晶胞 （2）基本的晶体结构 （3）晶面和米勒指数 （4）金刚石结构 （1）primitive and unit cell （2）basic crystal structures （3）crystal planes and Miller indices （4）diamond structure	0.5
1.4	原子价键 atomic bonding	原子价键 atomic bonding	
1.5	固体中的缺陷和杂质 imperfections and impurities in solids	（1）固体中的缺陷 （2）固体中的杂质 （1）imperfections in solids （2）impurities in solids	0.5
1.6	半导体材料的生长 growth of semiconductor materials	（1）在熔融体中生长 （2）外延生长 （1）growth from a melt （2）epitaxial growth	

第二章　量子力学初步（introduction to quantum mechanics）

章节序号 chapter number	章节名称 chapters	知识点 key points	课时 class hour
2.1	量子力学的基本原理 principles of quantum mechanics	（1）能量量子化 （2）波粒二相性 （3）不确定原理 （1）energy quanta （2）wave-particle duality （3）uncertainty principle	0.5
2.2	薛定谔波动方程 Schrodinger's wave equation	（1）波动方程 （2）波函数的物理意义 （3）边界条件	0.5

续表

章节序号 chapter number	章节名称 chapters	知识点 key points	课时 class hour
2.2	薛定谔波动方程 Schrodinger's wave equation	（1）wave equation （2）physical meaning of the wave function （3）boundary conditions	0.5
2.3	薛定谔波动方程的应用 applications of Schrodinger's wave equation	（1）自由空间中的电子 （2）无限深势阱 （3）阶跃势函数 （4）势垒和隧道效应 （1）electron in free space （2）infinite potential well （3）step potential function （4）potential barrier and tunneling	0.5
2.4	原子波动理论的延伸 extensions of the wave theory to atoms	（1）单电子原子 （2）周期表 （1）one-electron atom （2）periodic table	

第三章 固体量子理论初步（introduction to the quantum theory of solids）

章节序号 chapter number	章节名称 chapters	知识点 key points	课时 class hour
3.1	允带和禁带 allowed and forbidden energy bands	（1）能带的形成 （2）克龙尼克-潘纳模型 （3）k空间能带图 （1）formation of energy bands （2）Kronig-Penney model （3）k-space diagram	0.5
3.2	固体中电的传导 electrical conduction in solids	（1）能带和键模型 （2）漂移电流 （3）电子的有效质量 （4）空穴的概念 （5）金属、绝缘体和半导体 （1）energy band and the bond model （2）drift current （3）effective mass of electron （4）concept of the hole （5）metals, insulators, and semiconductors	1

续表

章节序号 chapter number	章节名称 chapters	知识点 key points	课时 class hour
3.3	三维扩展 extension to three dimensions	（1）硅和砷化镓的k空间能带图 （2）有效质量的补充概念 （1）k-space diagrams of Si and GaAs （2）additional effective mass concepts	0.5
3.4	状态密度函数 density of states function	（1）数学推导 （2）扩展到半导体 （1）mathematical derivation （2）extension to semiconductors	0.5
3.5	统计力学 statistical mechanics	（1）统计规律 （2）费米-狄拉克概率函数 （3）分布函数和费米能级 （1）statistical laws （2）Fermi-Dirac probability function （3）distribution function and the Fermi energy level	0.5

第四章　平衡半导体（semiconductor in equilibrium）

章节序号 chapter number	章节名称 chapters	知识点 key points	课时 class hour
4.1	半导体中的载流子 charge carriers in semiconductors	（1）电子和空穴的平衡分布 （2）n_0方程和p_0方程 （3）本征载流子浓度 （4）本征费米能级位置 （1）equilibrium distribution of electrons and holes （2）n_0 and p_0 equations （3）intrinsic carrier concentration （4）intrinsic Fermi energy level position	0.5
4.2	掺杂原子与能级 dopant atoms and energy levels	（1）定性描述 （2）电离能 （3）III-V族半导体 （1）qualitative description （2）ionization energy （3）group III-V semiconductors	0.5
4.3	非本征半导体 extrinsic semiconductor	（1）电子和空穴的平衡状态分布 （2）n_0和p_0的乘积 （3）费米-狄拉克积分 （4）简并和非简并半导体	0.5

续表

章节序号 chapter number	章节名称 chapters	知识点 key points	课时 class hour
4.3	非本征半导体 extrinsic semiconductor	（1）equilibrium distribution of electrons and holes （2）the $n_0 p_0$ product （3）Fermi-Dirac integral （4）degenerate and nondegenerate semiconductors	0.5
4.4	施主和受主的统计学分布 statistics of donors and acceptors	（1）概率分布函数 （2）完全电离和束缚态 （1）probability function （2）complete ionization and freeze-out	0.5
4.5	电中性状态 charge neutrality	（1）补偿半导体 （2）平衡电子和空穴浓度 （1）compensated semiconductors （2）equilibrium electron and hole concentrations	0.5
4.6	费米能级的位置 position of Fermi energy level	（1）数学推导 （2）E_F随掺杂浓度和温度的变化 （3）费米能级的应用 （1）mathematical derivation （2）variation of E_F with doping concentration and temperature （3）applications of the Fermi energy level	0.5

第五章 载流子输运现象（carrier transport phenomena）

章节序号 chapter number	章节名称 chapters	知识点 key points	课时 class hour
5.1	载流子的漂移运动 carrier drift	（1）飘逸电流密度 （2）迁移率 （3）电导率 （4）饱和速度 （1）drift current density （2）mobility effects （3）conductivity （4）velocity saturation	1
5.2	载流子扩散 carrier diffusion	（1）扩散电流密度 （2）总电流密度 （1）diffusion current density （2）total current density	1

续表

章节序号 chapter number	章节名称 chapters	知识点 key points	课时 class hour
5.3	杂质梯度分布 graded impurity distribution	（1）感生电场 （2）爱因斯坦关系 （1）induced electric field （2）Einstein relation	0.5
5.4	霍尔效应 Hall effect	霍尔效应 Hall effect	0.5

第六章　半导体中的非平衡过剩载流子（nonequilibrium excess carriers in semiconductors）

章节序号 chapter number	章节名称 chapters	知识点 key points	课时 class hour
6.1	载流子的产生与复合 generation and recombination of the excess carrier	（1）平衡状态半导体 （2）过剩载流子的产生与复合 （1）semiconductor in equilibrium （2）generation and recombination of the excess carrier	0.5
6.2	过剩载流子的性质 characteristics of excess carriers	（1）连续性方程 （2）与时间有关的扩散方程 （1）continuity equations （2）time-dependent diffusion equations	0.5
6.3	双极输运 ambipolar transport	（1）双极输运方程的推导 （2）掺杂与小注入的约束条件 （3）双极输运方程的应用 （4）介电弛豫时间常数 （5）海恩斯-肖克莱实验 （1）derivation of the ambipolar transport equation （2）limits of doping and low injection （3）applications of the ambipolar transport equation （4）dielectric relaxation time constant （5）Haynes-Shockley experiment	0.5
6.4	准费米能级 quasi-Fermi level	准费米能级 quasi-Fermi level	0.5
6.5	过剩载流子的寿命 excess-carrier lifetime	（1）肖克莱-里德-霍尔复合理论 （2）非本征掺杂和小注入的约束条件 （1）Shockley-Read-Hall theory of recombination （2）limits of extrinsic doping and low injection	0.5

续表

章节序号 chapter number	章节名称 chapters	知识点 key points	课时 class hour
6.6	表面效应 surface effects	（1）表面态 （2）表面复合速度 （1）surface states （2）surface recombination velocity	0.5

第七章　PN 结（PN junction）

章节序号 chapter number	章节名称 chapters	知识点 key points	课时 class hour
7.1	PN结的基本结构 basic structure of the PN junction	PN结的基本结构 basic structure of the PN junction	0.5
7.2	零偏 zero applied bias	（1）内建电势差 （2）电场强度 （3）空间电荷区宽度 （1）built-in potential barrier （2）electric field （3）space charge width	1
7.3	反偏 reverse applied bias	（1）空间电荷区宽度与电场 （2）结电容 （3）单边突变结 （1）space charge width and electric field （2）junction capacitance （3）one-sided junctions	1
7.4	非均匀掺杂PN结 nonuniformly doped PN junctions	（1）线性缓变结 （2）超突变结 （1）linearly graded junctions （2）hyperabrupt junctions	0.5

第八章　PN 结二极管（PN junction diode）

章节序号 chapter number	章节名称 chapters	知识点 key points	课时 class hour
8.1	PN结电流 PN junction current	（1）PN结内电荷流动的定性描述 （2）理想的电流–电压关系 （3）边界条件	1.5

章节序号 chapter number	章节名称 chapters	知识点 key points	课时 class hour
8.1	PN结电流 PN junction current	（4）少数载流子分布 （5）理想PN结电流 （6）物理学小结 （7）温度效应 （8）短二极管 （1）qualitative description of charge flow in a PN junction （2）ideal current-voltage relationship （3）boundary conditions （4）minority carrier distribution （5）ideal PN junction current （6）summary of physics （7）temperature effects （8）the "short" diode	1.5
8.2	PN结的小信号模型 small-signal model of the PN junction	（1）扩散电阻 （2）小信号导纳 （3）等效电路 （1）diffusion resistance （2）small-signal admittance （3）equivalent circuit	0.5
8.3	产生和复合电流 generation-recombination currents	（1）反偏产生电流 （2）正偏复合电流 （3）总正偏电流 （1）reverse-biased generation current （2）forward-bias recombination current （3）total forward-bias current	0.5
8.4	结击穿 junction breakdown	结击穿 junction breakdown	
8.5	电荷存储与二极管瞬态 charge storage and diode transients	（1）关瞬态 （2）开瞬态 （1）turn-off transient （2）turn-on transient	0.5
8.6	隧道二极管 tunnel diode	隧道二极管 tunnel diode	

第九章 金属 - 半导体和半导体异质结（metal-semiconductor and semiconductor heterojunctions）

章节序号 chapter number	章节名称 chapters	知识点 key points	课时 class hour
9.1	肖特基势垒二极管 Schottky barrier diode	（1）性质上的特征 （2）理想结的特征 （3）影响肖特基势垒高度的非理想因素 （4）电流-电压关系 （5）肖特基势垒二极管与PN结二极管的比较 （1）qualitative characteristics （2）ideal junction properties （3）nonideal effects on the Schottky barrier height （4）current-voltage relationship （5）comparison of the Schottky barrier diode and the PN junction diode	1.5
9.2	金属-半导体的欧姆接触 metal-semiconductor Ohmic contacts	（1）理想非整流接触势垒 （2）隧道效应 （3）接触电阻 （1）ideal nonrectifying barrier （2）tunneling effect （3）specifific contact resistance	0.5
9.3	异质结 heterojunctions	（1）形成异质结的材料 （2）能带图 （3）二维电子气 （4）静电平衡态 （5）I-V特性 （1）heterojunction materials （2）energy-band diagrams （3）two-dimensional electron gas （4）equilibrium electrostatics （5）current-voltage characteristics	1

第十章 双极晶体管（bipolar transistor）

章节序号 chapter number	章节名称 chapters	知识点 key points	课时 class hour
10.1	双极晶体管的工作原理 principle of bipolar transistor	（1）基本工作原理 （2）晶体管电流的简化表达式 （3）工作模式 （4）双极晶体管放大电路	0.5

续表

章节序号 chapter number	章节名称 chapters	知识点 key points	课时 class hour
10.1	双极晶体管的工作原理 principle of bipolar transistor	（1）basic principle of operation （2）simplified transistor current relation （3）modes of operation （4）amplification with bipolar transistors	0.5
10.2	少子的分布 minority carrier distribution	（1）正向有源模式 （2）其他工作模式 （1）forward-active mode （2）other modes of operation	
10.3	低频共基极电流增益 low-frequency common-base current gain	（1）有用的因素 （2）电流增益的数学表达式 （3）小结 （4）电流增益的计算 （1）contributing factors （2）derivation of current gain factors （3）summary （4）calculation examples of the gain factors	0.5
10.4	非理想效应 nonideal effects	（1）基区宽度调制效应 （2）大注入效应 （3）发射区禁带变窄 （4）电流集边效应 （5）基区非均匀掺杂的影响 （6）击穿电压 （1）base width modulation （2）high injection （3）emitter bandgap narrowing （4）current crowding （5）nonuniform base doping （6）breakdown voltage	1
10.5	等效电路模型 equivalent circuit models	（1）埃伯斯-莫尔模型 （2）根梅尔-普恩模型 （3）H-P模型 （1）Ebers-Moll model （2）Gummel-Poon model （3）Hybrid-Pi model	0.5
10.6	频率上限 frequency upper limitations	（1）延时因子 （2）晶体管截止频率 （1）time-delay factors （2）transistor cutoff frequency	

续表

章节序号 chapter number	章节名称 chapters	知识点 key points	课时 class hour
10.7	大信号开关 large-signal switching	（1）开关特性 （2）肖特基钳位晶体管 （1）switching characteristics （2）Schottky-clamped transistor	0.5
10.8	其他的双极晶体管结构 other bipolar transistor structures	（1）多晶硅发射区双极结型晶体管 （2）SiGe基区晶体管 （3）异质结双极晶体管 （1）polysilicon emitter BJT （2）Silicon-Germanium base transistor （3）heterojunction bipolar transistors	

第十一章 金属－氧化物－半导体场效应晶体管基础（fundamentals of the metal-oxide-semiconductor field-effect transistor）

章节序号 chapter number	章节名称 chapters	知识点 key points	课时 class hour
11.1	双端MOS结构 two-terminal MOS structure	（1）能带图 （2）耗尽层厚度 （3）功函数差 （4）平带电压 （5）阈值电压 （6）表面电荷分布 （1）energy-band diagrams （2）depletion layer thickness （3）work function differences （4）flat-band voltage （5）threshold voltage （6）surface charge density	1
11.2	电容-电压特性 capacitance-voltage characteristics	（1）理想C-V特性 （2）频率特性 （3）固定栅氧化层电荷和界面电荷效应 （1）ideal C-V characteristics （2）frequency effects （3）fixed oxide and interface charge effects	0.5
11.3	MOSFET基本工作原理 basic MOSFET operation	（1）MOSFET结构 （2）电流-电压关系的概念 （3）电流-电压关系的数学推导	1

续表

章节序号 chapter number	章节名称 chapters	知识点 key points	课时 class hour
11.3	MOSFET基本工作原理 basic MOSFET operation	（4）跨导 （5）衬底偏置效应 （1）MOSFET structures （2）concept of current-voltage relationship （3）mathematical derivation of current-voltage relationship （4）transconductance （5）substrate bias effects	1
11.4	频率限制特性 frequency limitations	（1）小信号等效电路 （2）频率限制因素与截止频率 （1）small-signal equivalent circuit （2）frequency limitation factors and cutoff frequency	0.5
11.5	CMOS技术 CMOS technology	CMOS技术 CMOS technology	

第十二章 金属－氧化物－半导体场效应晶体管：概念的深入（metal-oxide-semiconductor field-effect transistor: additional concepts）

章节序号 chapter number	章节名称 chapters	知识点 key points	课时 class hour
12.1	非理想效应 nonideal effects	（1）亚阈值电导 （2）沟道长度调制效应 （3）迁移率变化 （4）速度饱和 （5）弹道输运 （1）subthreshold conduction （2）channel length modulation （3）mobility variation （4）velocity saturation （5）ballistic transport	1
12.2	MOSFET按比例缩小理论 MOSFET scaling law	（1）恒定电场按比例缩小 （2）阈值电压——一级近似 （3）全部按比例缩小理论 （1）constant-field scaling （2）threshold voltage—first approximation （3）generalized scaling	0.5

续表

章节序号 chapter number	章节名称 chapters	知识点 key points	课时 class hour
12.3	阈值电压的修正 threshold voltage modifications	（1）短沟道效应 （2）窄沟道效应 （1）short-channel effects （2）narrow-channel effects	0.5
12.4	附加电学特性 additional electrical characteristics	（1）击穿电压 （2）轻掺杂漏晶体管 （3）通过离子注入进行阈值调整 （1）breakdown voltage （2）lightly doped drain transistor （3）threshold adjustment by ion implantation	0.5
12.5	辐射和热电子效应 radiation and hot-electron effects	（1）辐射引入的氧化层电荷 （2）辐射引入的界面态 （3）热电子充电效应 （1）radiation-induced oxide charge （2）radiation-induced interface states （3）hot-electron charging effects	0.5

第十三章 结型场效应晶体管介绍（junction field-effect transistor）

章节序号 chapter number	章节名称 chapters	知识点 key points	课时 class hour
13.1	JFET的概念 concept of JFET	（1）PN JFET的基本工作原理 （2）MESFET的基本工作原理 （1）basic PN JFET operation （2）basic MESFET operation	
13.2	器件的特性 characteristics of devices	（1）内建夹断电压、夹断电压和漏源饱和电压 （2）耗尽型JFET的理想直流I-V特性 （3）跨导 （4）MESFET （1）internal pinchoff voltage, pinchoff voltage, and drain-to-source saturation voltage （2）ideal DC current-voltage relationship—depletion mode JFET （3）transconductance （4）MESFET	0.5
13.3	非理想效应 nonideal effects	（1）沟道长度调制效应 （2）饱和速度影响	0.5

续表

章节序号 chapter number	章节名称 chapters	知识点 key points	课时 class hour
13.3	非理想效应 nonideal effects	（3）亚阈值特性和栅电流效应 （1）channel length modulation （2）velocity saturation effects （3）subthreshold and gate current effects	0.5
13.4	等效电路和频率限制 equivalent circuit and frequency limitations	（1）小信号等效电路 （2）频率限制因子和截止频率 （1）small-signal equivalent circuit （2）frequency limitation factors and cutoff frequency	0.5
13.5	高电子迁移率晶体管 high electron mobility transistor	（1）量子阱结构 （2）晶体管性能 （1）quantum well structures （2）transistor performance	

第十四章 光器件（optical devices）

章节序号 chapter number	章节名称 chapters	知识点 key points	课时 class hour
14.1	光学吸收 optical absorption	（1）光子吸收系数 （2）电子-空穴对的产生率 （1）photon absorption coefficient （2）electron-hole pair generation rate	0.5
14.2	太阳能电池 solar cells	（1）PN结太阳能电池 （2）转换效率和太阳能集中 （3）非均匀吸收的影响 （4）异质结太阳能电池 （5）非晶态（无定型）硅太阳能电池 （1）PN junction solar cell （2）conversion effifiency and solar concentration （3）nonuniform absorption effects （4）heterojunction solar cell （5）amorphous silicon solar cells	0.5
14.3	光电探测器 photodetectors	（1）光电导体 （2）光电二极管 （3）PIN光电二极管 （4）雪崩光电二极管 （5）光电晶体管 （1）photoconductor	0.5

章节序号 chapter number	章节名称 chapters	知识点 key points	课时 class hour
14.3	光电探测器 photodetectors	（2）photodiode （3）PIN photodiode （4）avalanche photodiode （5）phototransistor	0.5
14.4	光致发光和电致发光 photoluminescence and electroluminescence	（1）基本跃迁 （2）发光效率 （3）材料 （1）basic transitions （2）luminescent efficiency （3）materials	0.5
14.5	光电二极管 light emitting diodes	（1）光的产生 （2）内量子效率 （3）外量子效率 （4）LED器件 （1）generation of light （2）internal quantum efficiency （3）external quantum efficiency （4）LED devices	0.5
14.6	激光二极管 laser diodes	（1）受激辐射和分布反转 （2）光学空腔谐振器 （3）阈值电流 （4）器件结构与特性 （1）stimulated emission and population inversion （2）optical cavity （3）threshold current （4）structures and characteristics of devices	0.5

第十五章 半导体功率器件（semiconductor power devices）

章节序号 chapter number	章节名称 chapters	知识点 key points	课时 class hour
15.1	功率双极晶体管 power bipolar transistors	（1）垂直式功率晶体管的结构 （2）功率晶体管的特性 （3）达林顿组态 （1）structure of vertical power transistor （2）characteristics of power transistor （3）Darlington pair configuration	0.5

续表

章节序号 chapter number	章节名称 chapters	知识点 key points	课时 class hour
15.2	功率MOSFET power MOSFET	（1）功率晶体管的结构 （2）功率MOSFET的特性 （3）寄生双极晶体管 （1）structures of power transistor （2）characteristics of power MOSFET （3）parasitic BJT	0.5
15.3	散热片和结温 heat sinks and junction temperature	散热片和结温 heat sinks and junction temperature	
15.4	半导体闸流管 thyristor	（1）半导体闸流管的基本特性 （2）SCR的触发机理 （3）SCR的关断 （4）器件结构 （1）basic characteristics of thyristor （2）triggering the SCR （3）SCR turn-off （4）structures of devices	0.5

6.3.5　实验环节（Experiments）

序号 number	实验内容 experiment content	知识点 key points	课时 class hour
1	四探针法测半导体电阻率 four-probe method to measure semiconductor resistivity	（1）对材料和样品分别测量其体电阻率和薄层电阻 R_s （2）分析测量电阻率和薄层电阻中产生误差的来源 （1）measure volumetric resistivity and sheet resistance R_s of materials and samples respectively （2）analyze the source of errors in measuring resistivity and sheet resistance	4
2	采用霍尔效应测试半导体载流子特性和浓度 using Hall effect to test semiconductor carrier characteristics and concentration	（1）搭建测试电路 （2）对霍尔样品完成电阻率测量、霍尔系数测量 （3）通过计算机编程，根据测量数据，确定样品导电类型 （4）计算出霍尔样品的载流子浓度及霍尔迁移率，对结果进行定量分析	4

序号 number	实验内容 experiment content	知识点 key points	课时 class hour
2	采用霍尔效应测试半导体载流子特性和浓度 using Hall effect to test semiconductor carrier characteristics and concentration	（1）build the circuit （2）complete resistivity measurement and Hall coefficient measurement for Hall samples （3）through computer programming, determine the sample conductivity type according to the measurement data （4）calculate the carrier concentration and Hall mobility of the Hall sample, and quantitatively analyze the results	4
3	二极管特性测试 diode characteristic test	（1）测量二极管反向恢复时间和反向恢复电荷并研究测试条件变化对测试值的影响 （2）记录在双踪示波器上观察到的相关波形 （3）记录在一定测试条件下二极管的反向恢复时间和反向恢复电荷 （4）改变测试条件，对测试结果进行分析 （1）measure diode reverse recovery time and reverse recovery charge and study the effect of test conditions on test values （2）record related waveforms observed on a dual-track oscilloscope （3）record the reverse recovery time and reverse recovery charge of the diode under certain test conditions （4）change test conditions and analyze test results	4
4	三极管特性测试 transistor characteristic test	（1）掌握晶体管特征频率与器件结构、偏置电压等参数的变化关系 （2）熟悉测试方法，利用测试所得的实验数据分别画出特征频率与 V_{ce}、I_e、$1/I_e$ 等的关系曲线 （3）对相应的曲线和测量数据进行分析讨论 （1）master the relationship between transistor characteristic frequency and device structure, bias voltage, and other parameters （2）acquainted with the test method, use the experimental data obtained by the test to draw the relationship curve between the characteristic frequency and V_{ce}, I_e, $1/I_e$, etc. （3）analyze and discuss the corresponding curve and measurement data	4

6.4　"电池材料制备技术"教学大纲

课程名称：电池材料制备技术

Course Title：Battery Materials Preparation Technology

先修课程：储能化学基础（含实验），储能材料工程

Prerequisites：General Chemistry for Energy Storage（with Experiments），Energy Storage Material Engineering

学分：2

Credits：2

6.4.1　课程目的和基本内容（Course Objectives and Basic Contents）

本课程是储能科学与工程专业的一门选修课。

本课程围绕材料制备技术，为学生详细讲授不同制备方法，包括液相法、固相法、气相法等的基本概念、特点及应用要求，并介绍各种制备方法在不同储能体系（二次电池、太阳能电池、燃料电池等）中的应用。结合具体材料的制备及分析，使学生通过学习基础知识准确理解材料制备的基本核心概念及设计理念，培养学生利用理论知识解决实际技术问题的能力。

课程教学中，既要讲授储能体系的基础理论知识，还要通过具体电池材料设计制备实例分析，使学生在掌握材料制备技术的同时，理解不同制备技术所获得的电池材料性质与性能之间的构效关系，为特性体系电池材料的设计、制备及改性提供理论基础。

This course is an elective course for undergraduates majoring in energy storage science and engineering.

This course focuses on material preparation technology and introduces the main preparation methods of battery materials, including liquid phase method, solid phase method, and gas phase method. It also presents various preparation methods in different energy storage systems (secondary batteries, solar cells, fuel cells, etc.). The relevant knowledge in this course provides an indispensable foundation of preparation methods for battery materials. Combined with the preparation and analysis of specific materials, students can learn basic knowledge, deeply understand the concepts and design of materials preparation, and promote their ability to solve practical technical problems.

This course focuses on the theoretical lectures to teach the basic knowledge of energy storage systems and preparation methods of battery materials. Through the study of this course, students can not only master the theoretical knowledge and concepts of material preparation technology, but also understand the structure-activity relationship of battery materials obtained by different preparation technologies. It is beneficial for students to lay a solid theoretical basis for the design, preparation, and modification of battery materials.

6.4.2 课程基本情况（Basic Information of the Course）

课程名称	电池材料制备技术 Battery Materials Preparation Technology								
开课时间	一年级		二年级		三年级		四年级		总学分	2	总学时	40
	秋	春	秋	春	秋	春	秋	春				
课程定位	储能科学与工程专业本科生电化学与电磁储能课程群（B模块）选修课											
授课学时分配	课堂讲授24学时+实验16学时											
先修课程	储能化学基础（含实验),储能材料工程											
后续课程	无											
教学方式	课堂教学,实验与综述报告,自学报告											
考核方式	课程结束笔试成绩占60%,实验报告占20%,平时成绩占20%											
参考教材	[1] 张以河.材料制备化学.北京:化学工业出版社,2013.											
参考资料	[1] 黄志高,林应斌,李传常. 储能原理与技术. 北京:中国水利水电出版社,2018. [2] 张现发.高性能锂离子电池电极材料的制备与性能研究.哈尔滨:黑龙江大学出版社,2018. [3] 王红强.应用化学综合实验.北京:化学工业出版社,2019.											

6.4.3 教学目的和基本要求（Teaching Objectives and Basic Requirements）

（1）具备电池材料制备技术基础知识理解能力。掌握电池材料制备技术的基础知识，包括储能器件、材料制备技术、材料设计理念及性能指标，熟悉各类制备技术（液相法、固相法、气相法等）的原理、特征及应用要求，掌握常用电池材料的制备方法。

（2）具备电池材料的分析与设计能力。根据不同电池材料的特点，能够运用电池材料制备技术的相关知识，对不同体系、不同类型的电池材料提出合理可行的制备途径，并能通过材料表征技术准确地分析电池材料的合成方法对材料性质的影响。

（1）Understand the basic knowledge of battery materials preparation technology. In this course, students should master the basic knowledge of battery materials preparation technology, including energy storage devices, materials preparation technology, materials design concepts, and performance indices. Besides, they can become familiar with the principles, characteristics, and application requirements of various preparation technologies (such as liquid phase method, solid phase method, gas phase method, etc.), and master the preparation methods of commonly used battery materials.

（2）Master the ability to analyze and design battery materials. Students can use the relevant knowledge of battery materials preparation technology to propose reasonable and feasible preparation methods for different systems and different types of battery materials based on the characteristics of battery materials. Furthermore, they can accurately analyze the effect of the synthesis method of battery materials on the material properties through materials characterization technology.

6.4.4　课程大纲和知识点（Syllabus and Key Points）

第一章　绪论（introduction）

章节序号 chapter number	章节名称 chapters	知识点 key points	课时 class hour
1.1	电池材料发展概况 progress of battery materials	（1）电池材料研究现状 （2）电池材料发展趋势 （1）research status of battery materials （2）trend of battery materials	1
1.2	本课程的任务 tasks of this course	（1）制备方法简介 （2）本课程的重难点 （1）introduction to the preparation method （2）key and difficult points of this course	1

第二章　液相法（liquid phase method）

章节序号 chapter number	章节名称 chapters	知识点 key points	课时 class hour
2.1	水热法 hydrothermal method	（1）水热法的基本概念及原理 （2）水热法的特征 （3）水热法的影响因素 （4）水热法的应用要求 （1）basic concepts and principles of hydrothermal method （2）characteristics of hydrothermal method （3）factors to influence hydrothermal method （4）application requirements of hydrothermal method	1
2.2	溶胶凝胶法 sol-gel method	（1）溶胶凝胶法的基本概念及原理 （2）溶胶凝胶法的特征 （3）溶胶凝胶法的影响因素 （4）溶胶凝法的应用要求	1.5

续表

章节序号 chapter number	章节名称 chapters	知识点 key points	课时 class hour
2.2	溶胶凝胶法 sol-gel method	（1）basic concepts and principles of sol-gel method （2）characteristics of sol-gel method （3）factors to influence sol-gel method （4）application requirements of sol-gel method	1.5
2.3	共沉淀法 co-precipitation method	（1）共沉淀法的基本概念及原理 （2）共沉淀法的特征 （3）共沉淀法的影响因素 （4）共沉淀法的应用要求 （1）basic concepts and principles of co-precipitation method （2）characteristics of co-precipitation method （3）factors to influence co-precipitation method （4）application requirements of co-precipitation method	1.5
2.4	模板法 template method	（1）模板法的基本概念及原理 （2）模板法的特征 （3）模板法的应用要求 （1）basic concepts and principles of template method （2）characteristics of template method （3）application requirements of template method	1
2.5	液相法在电池材料制备中的应用 liquid phase method in the preparation of battery materials	（1）液相法在锂离子电池材料中的应用 （2）液相法制备的锂离子电池正极材料举例 （1）application of liquid phase method in lithium-ion battery materials （2）examples of lithium-ion battery cathode materials prepared by liquid phase method	1

第三章　固相法（solid phase method）

章节序号 chapter number	章节名称 chapters	知识点 key points	课时 class hour
3.1	固相反应法 solid state reaction method	（1）固相反应法的基本概念及原理 （2）固相反应法特征 （3）固相反应法的影响因素 （4）固相反应法的应用要求	1.5

章节序号 chapter number	章节名称 chapters	知识点 key points	课时 class hour
3.1	固相反应法 solid state reaction method	（1）basic concepts and principles of solid state reaction method （2）characteristics of solid state reaction method （3）factors to influence solid state reaction method （4）application requirements of solid state reaction method	1.5
3.2	热分解法 thermal decomposition method	（1）热分解法的基本概念及原理 （2）热分解法的特征 （3）热分解法的应用要求 （1）basic concepts and principles of thermal decomposition method （2）characteristics of thermal decomposition method （3）application requirements of thermal decomposition method	1
3.3	球磨法 ball milling method	（1）球磨法的基本概念及原理 （2）球磨法的特征 （3）球磨法的影响因素 （4）球磨法的应用要求 （1）basic concepts and principles of ball milling method （2）characteristics of ball milling method （3）factors to influence ball milling method （4）application requirements of ball milling method	1.5
3.4	溶出法 dissolution method	（1）溶出法的基本概念及原理 （2）溶出法的特征 （3）溶出法的应用要求 （1）basic concepts and principles of dissolution method （2）characteristics of dissolution method （3）application requirements of dissolution method	1
3.5	固相法在电池材料制备中的应用 solid phase method in the preparation of battery materials	（1）固相法在电池材料中的应用现状 （2）固相法制备的电池材料举例 （1）application situation of solid phase method in battery materials （2）examples of battery materials prepared by solid phase method	1

第四章　气相法（gas phase method）

章节序号 chapter number	章节名称 chapters	知识点 key points	课时 class hour
4.1	冷凝法 condensation method	（1）冷凝法的基本概念及原理 （2）冷凝法的特征 （3）冷凝法的影响因素 （4）冷凝法的应用要求 （1）basic concepts and principles of condensation method （2）characteristics of condensation method （3）factors to influence condensation method （4）application requirements of condensation method	1.5
4.2	激光消融法 laser ablation	（1）激光消融法的基本概念及原理 （2）激光消融法的特征 （3）激光消融法的应用要求 （1）basic concepts and principles of laser ablation （2）characteristics of laser ablation （3）application requirements of laser ablation	1
4.3	蒸发法 evaporation method	（1）蒸发法的基本概念及原理 （2）蒸发法的特征 （3）蒸发法的应用要求 （1）basic concepts and principles of evaporation method （2）characteristics of evaporation method （3）application requirements of evaporation method	1
4.4	溅射法 sputtering method	（1）溅射法的基本概念及原理 （2）溅射法的特征 （3）溅射法的影响因素 （4）溅射法的应用要求 （1）basic concepts and principles of sputtering method （2）characteristics of sputtering method （3）factors to influence sputtering method （4）application requirements of sputtering method	1.5
4.5	气相法在电池材料制备中的应用 gas phase method in the preparation of battery materials	（1）气相法在电池材料中的应用现状 （2）气相法制备的电池材料举例 （1）application situation of gas phase method in battery materials （2）examples of battery materials prepared by gas phase method	1

第五章　其他电池材料制备方法（other preparation methods of battery materials）

章节序号 chapter number	章节名称 chapters	知识点 key points	课时 class hour
5.1	微波辐射法 microwave radiation method	（1）微波辐射法的基本概念及原理 （2）微波辐射法的特征 （3）微波辐射法的应用要求 （1）basic concepts and principles of microwave radiation method （2）characteristics of microwave radiation method （3）application requirements of microwave radiation method	1
5.2	剥离法 stripping method	（1）剥离法的基本概念及原理 （2）剥离法的特征 （3）剥离法的应用要求 （1）basic concepts and principles of stripping method （2）characteristics of stripping method （3）application requirements of stripping method	1
5.3	压淬法 press quenching method	（1）压淬法的基本概念及原理 （2）压淬法的特征 （3）压淬法的应用要求 （1）basic concepts and principles of press quenching method （2）characteristics of press quenching method （3）application requirements of press quenching method	0.5
5.4	脉冲电流法 pulse current method	（1）脉冲电流法的基本概念及原理 （2）脉冲电流法的特征 （3）脉冲电流法的应用要求 （1）basic concepts and principles of pulse current method （2）characteristics of pulse current method （3）application requirements of pulse current method	0.5
5.5	其他方法在电池材料制备中的应用 other methods in the preparation of battery materials	（1）其他方法在电池材料中的应用现状 （2）其他方法制备的电池材料举例 （1）application situation of other methods in battery materials （2）examples of battery materials prepared by other methods	1

6.4.5 实验环节（Experiments）

序号 number	实验内容 experiment content	知识点 key points	课时 class hour
1	共沉淀法制备三元正极材料 preparation of ternary cathode by co-precipitation method	（1）熟悉实验操作流程 （2）材料制备 （3）材料成分表征与分析 （1）acquaint with the experimental process （2）prepare electrode material （3）test and evaluate the compositions of the material	4
2	溶胶凝胶法制备磷酸铁锂正极材料 preparation of lithium iron phosphate cathode by sol-gel method	（1）熟悉实验操作流程 （2）材料制备 （3）材料成分表征与分析 （1）acquaint with the experimental process （2）prepare electrode material （3）test and evaluate the compositions of the material	4
3	球磨法制备钴酸锂正极材料 preparation of lithium cobalt oxide cathode by ball milling	（1）熟悉实验操作流程 （2）材料制备 （3）材料成分表征与分析 （1）acquaint with the experimental process （2）prepare electrode material （3）test and evaluate the compositions of the material	4
4	溅射法改性正极材料 surface modification of cathode by sputtering	（1）熟悉实验操作流程 （2）用溅射法对正极材料表面改性 （3）材料表征与分析 （1）acquaint with the experimental process （2）modify the surface of cathode by sputtering （3）test and analyze the material	4

6.5 "新型储能电池技术"教学大纲

课程名称：新型储能电池技术

Course Title：Advanced Energy Storage Technologies

先修课程：电化学基础，储能材料工程

Prerequisites：Fundamentals of Electrochemistry，Energy Storage Materials Engineering

学分：2

Credits：2

6.5.1 课程目的和基本内容（Course Objectives and Basic Contents）

本课程是储能科学与工程专业的一门专业选修课。

本课程以先进电池储能材料与技术为核心，为学生详细讲授了储能电池电化学的基础知识、储能电池技术的发展现状、几种新型的液流电池、高温钠硫电池、ZEBRA 电池、液态金属电池、钠离子电池的基本机理和电极材料发展现状及目前遇到的挑战，重点就新型储能电池技术中的电极材料和电解质的选择设计、电池的结构设计以及其储能特性和应用等方面进行介绍，使学生基本掌握多种储能技术的储能机理和前沿发展动态，并使学生能够从具体储能场景及要求出发，选择相应的储能材料及技术，锻炼学生理论结合实际的能力。

This course is an elective course for undergraduates majoring in Energy Storage Science and Engineering.

This course mainly focuses on advanced battery technologies for energy storage. Here, we will provide students an overview of energy storage materials and technologies. Topics include fundamentals of energy chemistry, the principle and current research status of a variety of energy storage technologies such as redox flow battery, sodium-sulfur battery, ZEBRA battery, liquid metal battery, and sodium-ion battery. And the challenge of energy storage technologies will be also discussed. We will focus on the introduction of the selection rules of electrodes and electrolytes for different battery technologies, the structure design of the battery, the storage characteristics, and its application. We will make the students understand the principle of different battery technologies and their current research status. The students can design the energy storage scenario by choosing the appropriate technology according to the application.

6.5.2 课程基本情况（Basic Information of the Course）

课程名称	新型储能电池技术 Advanced Energy Storage Technologies											
开课时间	一年级		二年级		三年级		四年级		总学分	2	总学时	40
	秋	春	秋	春	秋	春	秋	春				
课程定位	储能科学与工程专业本科生电化学与电磁储能课程群（B模块）选修课											
授课学时分配	课堂讲授24学时+实验16学时											
先修课程	电化学基础,半导体物理,储能材料工程											
后续课程	无											
教学方式	课堂教学,大作业,实验与综述报告,平时作业											
考核方式	课程结束笔试成绩占70%,实验报告占20%,平时成绩占10%											

参考教材	[1] Huggins R A. Advanced Batteries: Materials Science Aspects. Berlin：Springer, 2008. [2] 丁玉龙，来小康，陈海生.储能技术及应用.北京：化学工业出版社，2019.
参考资料	[1] 张华民. 液流电池技术.北京：化学工业出版社，2014. [2] 杨勇. 固态电化学. 北京：化学工业出版社，2016.

6.5.3　教学目的和基本要求（Teaching Objectives and Basic Requirements）

通过本课程的学习，使学生掌握各种储能电池技术的基本知识，并为之后的学习打下良好的基础。

基本要求：

（1）电池电化学储能技术领域基础知识理解能力。掌握电池电化学的基础知识，熟悉各类储能材料在各种储能技术中（包括液流电池、钠硫电池、ZEBRA 电池、液态金属电池和钠离子电池）的发展现状和面临的挑战及问题，并能够详细了解各类储能技术的储能特性、应用场景和不足。

（2）电池材料与储能技术的设计与创新能力。根据储能场景的具体要求，选择合适的储能技术并选择相应的储能材料，在材料选择和储能技术设计中体现创新意识。

Through the study of this course, students can obtain the basic knowledge and skills of advanced battery technology, and lay the foundation for the subsequent study of related courses.

Basic requirements:

（1）Understand the fundamental of energy chemistry, the principle and current research status of a variety of energy storage technologies such as redox flow battery, sodium-sulfur battery, ZEBRA battery, liquid metal battery, and sodium-ion battery. And the status, challenges and current problems of the energy storage technologies are also required to master. Furthermore, the characteristics, application scenarios and shortcomings of various energy storage technologies have been known in detail.

（2）Innovation capability on the materials design for energy storage technology.Design the energy storage scenario by choosing the appropriate technology and materials according to the application with the creative ideas.

6.5.4　课程大纲和知识点（Syllabus and Key Points）

第一章　绪论（introduction）

章节序号 chapter number	章节名称 chapters	知识点 key points	课时 class hour
1.1	储能电池技术简介 introduction of battery technologies for energy storage	（1）风能、太阳能可再生能源的发展现状 （2）储能电池技术在可再生能源发电中的作用 （1）state of the art of solar and wind energy （2）role of battery technology for energy storage in the application of renewable energy	1
1.2	储能电池电化学基础 fundamental of electrochemistry for energy storage battery	（1）一次电池 （2）二次电池 （3）电池的容量与能量 （1）primary battery （2）secondary battery （3）capacity and energy for battery	1

第二章　液流电池技术（redox flow battery）

章节序号 chapter number	章节名称 chapters	知识点 key points	课时 class hour
2.1	液流电池概述 introduction of redox flow battery	（1）液流电池的基本原理 （2）液流电池的发展现状及未来的挑战 （1）principle of redox flow battery （2）current research status and challenges of redox flow battery	2
2.2	几种典型液流电池的材料体系 typical redox flow battery systems	（1）全钒液流电池 （2）锌/溴液流电池 （1）vanadium redox flow battery （2）zinc-bromine redox flow battery	2

第三章　高温钠硫电池技术（sodium-sulfur battery）

章节序号 chapter number	章节名称 chapters	知识点 key points	课时 class hour
3.1	钠硫电池技术概述 introduction of sodium-sulfur battery	（1）钠硫电池的原理 （2）钠硫电池的特点 （3）钠硫电池的挑战	2

续表

章节序号 chapter number	章节名称 chapters	知识点 key points	课时 class hour
3.1	钠硫电池技术概述 introduction of sodium-sulfur battery	（1）working principle of sodium-sulfur battery （2）characteristic of sodium-sulfur battery （3）challenges of sodium-sulfur battery	2
3.2	钠硫电池的设计与应用 design and application of sodium-sulfur battery	（1）管式钠硫电池 （2）钠硫电池的应用 （3）新型钠硫电池的发展 （1）tubular sodium-sulfur battery （2）application of sodium-sulfur battery （3）development of new type sodium-sulfur battery	2

第四章　ZEBRA 电池技术（ZEBRA battery）

章节序号 chapter number	章节名称 chapters	知识点 key points	课时 class hour
4.1	ZEBRA电池概述 introduction of ZEBRA battery	（1）ZEBRA电池的原理 （2）ZEBRA电池的储能特性 （3）ZEBRA电池的发展和挑战 （1）working principle of ZEBRA battery （2）energy storage characteristics of ZEBRA battery （3）development and challenges of ZEBRA battery	2
4.2	ZEBRA电池的设计与应用 design and application of ZEBRA battery	（1）管式设计的ZEBRA电池 （2）平板式设计的ZEBRA电池 （1）tubular ZEBRA battery （2）tabulate ZEBRA battery	2

第五章　液态金属电池技术（liquid metal battery）

章节序号 chapter number	章节名称 chapters	知识点 key points	课时 class hour
5.1	液态金属电池概述 introduction of liquid metal battery	（1）液态金属电池的基本原理 （2）液态金属电池的结构 （3）液态金属电池的特点 （4）液态金属电池的发展和挑战 （1）working principle of liquid metal battery （2）structure of liquid metal battery （3）characteristics of liquid metal battery （4）development and challenges of liquid metal battery	2

章节序号 chapter number	章节名称 chapters	知识点 key points	课时 class hour
5.2	液态金属电池关键材料 key materials for liquid metal battery	（1）液态金属电池电极材料 （2）液态金属电池电解质材料 （1）electrode materials for liquid metal battery （2）electrolyte materials for liquid metal battery	2
5.3	几种典型的液态金属电池体系及其储能特性 several typical electrode and electrolyte systems for liquid metal battery and their energy storage characteristics	（1）Mg基液态金属电池 （2）Li基液态金属电池 （3）Ca基液态金属电池 （4）室温液态金属电池 （1）Mg-based liquid metal battery （2）Li-based liquid metal battery （3）Ca-based liquid metal battery （4）liquid metal battery at room temperature	2

第六章　钠离子电池（sodium-ion battery）

章节序号 chapter number	章节名称 chapters	知识点 key points	课时 class hour
6.1	钠离子电池的原理和特点 working principle and characteristics of sodium-ion battery	（1）钠离子电池的基本原理 （2）钠离子电池的价格因素 （3）钠离子电池在储能领域的发展 （1）working principle of sodium-ion battery （2）cost factor of sodium-ion battery （3）application of sodium-ion battery in grid energy storage	2
6.2	钠离子电池的关键材料 key materials for sodium-ion battery	（1）钠离子电池正极和负极材料的选择和设计思路 （2）水系钠离子电池 （1）selection and design of electrode materials in sodium-ion battery （2）aqueous sodium-ion battery	2

6.5.5 实验环节（Experiments）

序号 number	实验内容 experiment content	知识点 key points	课时 class hour
1	液态金属电池的设计和组装 design and assembling of liquid metal battery	（1）液态金属电池的结构设计 （2）液态金属电池的电极材料制备 （3）液态金属电池的电解质材料制备 （4）液态金属电池的组装 （1）structural design of liquid metal battery （2）preparation on electrodes for liquid metal battery （3）preparation on electrolyte for liquid metal battery （4）assembling of liquid metal battery	4
2	液态金属电池的储能特性测试 measurement on energy storage characteristics of liquid metal battery	（1）液态金属电池的温度控制 （2）液态金属电池的恒电流充放电测试 （3）液态金属电池的交流阻抗测试 （4）温度和电流密度对液态金属电池充放电性能的影响 （1）temperature control of liquid metal battery （2）galvanostatic charging-discharging test of liquid metal battery （3）AC impedance test of liquid metal battery （4）influence of temperature and current density on the electrochemical properties of liquid metal battery	4
3	钠离子电池的材料选择与电池组装 selection of electrode materials and assemble of sodium-ion battery	（1）钠离子电池的电极材料选择 （2）钠离子电池的电极材料制备 （3）钠离子电池的组装 （1）selection and design of electrode materials in sodium-ion battery （2）preparation on electrode materials for sodium-ion battery （3）assembling of sodium-ion battery	4
4	钠离子电池的特性测试 measurement on energy storage characteristics of sodium-ion battery	（1）钠离子电池的恒电流充放电测试 （2）钠离子电池的交流阻抗测试 （3）温度和电流密度对钠离子电池充放电性能的影响 （1）galvanostatic charging-discharging test of sodium-ion battery （2）AC impedance measurement of sodium-ion battery （3）influence of temperature and current density on the electrochemical properties of sodium-ion battery	4

6.6 "纳米材料与能源"教学大纲

课程名称：纳米材料与能源

Course Title：Nanomaterials and Energy

先修课程：电化学基础，半导体物理

Prerequisites：Fundamentals of Electrochemistry, Semiconductor Physics

学分：2

Credits：2

6.6.1 课程目的和基本内容（Course Objectives and Basic Contents）

本课程为储能科学与工程专业学生提供坚实的纳米材料的合成与应用基础。立足对纳米材料与能源本质的认识，系统地介绍了纳米材料的结构、性质、合成、表征及其在储能器件应用过程中的基本原理、基本理论，深入讨论纳米材料的物质组成、晶体结构与电化学储能性质间的关系，归纳、提炼出储能科学中最基础、最本质的原理及变化规律。

课程主要内容包括绪论、纳米材料的结构、性质及表征、纳米材料的制备方法（气相法、液相法、固相法）及其在储能器件（锂离子电池、化学电容器、燃料电池）中的应用。

The course aims to provide a basic knowledge of synthesis and application of nanomaterials for students majoring in energy storage science and engineering. It introduces the structure, composition, synthesis, and characterization of nanomaterials as well as their basic principles and theories in energy storage devices. The relationship between the composition, the crystal structure, and the electrochemical energy storage properties of nanomaterials is discussed deeply, so as to summarize and refined the most essential principles and change law in energy storage science.

The course includes the introduction, the structure, properties and characterization of nanomaterials, preparations of nanomaterials (such as chemical vappor deposition, liquid phase method, solid phase method), and their applications in energy storage devices (such as lithium-ion batteries, chemical capacitors, fuel cells).

6.6.2 课程基本情况（Basic Information of the Course）

课程名称	纳米材料与能源 Nanomaterials and Energy								
开课时间	一年级		二年级		三年级		四年级		总学分	2	总学时	40
	秋	春	秋	春	秋	春	秋	春				
课程定位	储能科学与工程专业本科生电化学与电磁储能课程群（B模块）选修课								
授课学时分配	课堂讲授24学时+实验16学时								
先修课程	电化学原理,半导体物理								
后续课程	无								
教学方式	课堂教学,实验与综述报告,讨论								
考核方式	平时成绩占30%（平时测验成绩占10%,平时作业占20%）,实验占20%,课程结束笔试成绩占50%								
参考教材	[1] 黄开金.纳米材料的制备与应用.北京:冶金工业出版社,2009. [2] Guo Zhanhu,Chen Yuan,Lu N L.多功能纳米复合材料及其在储能和环境中的应用（英文版）.北京:高等教育出版社,2018.								
参考资料	[1] 刘漫红,孙瑞雪,肖海连,等.纳米材料及其制备技术.北京:冶金工业出版社,2014. [2] 张凯峰.纳米材料成形理论与技术.哈尔滨:哈尔滨工业大学出版社,2012.								

6.6.3 教学目的和基本要求（Teaching Objectives and Basic Requirements）

（1）熟悉纳米材料合成的基本方法、基本原理和基本特点，理解纳米材料的组成、结构和化学储能性质的关系；并能应用这些理论解释纳米材料合成方法、化学储能性质的关系及其变化规律。

（2）掌握纳米材料的常用合成方法和基本原理，熟悉纳米材料在储能器件中的储能原理。

（3）熟悉锂离子电池的基本结构和工作原理，掌握锂离子电池正负极材料的制备方法，熟悉锂离子电池的常用表征技术。

（4）熟悉化学电容器的基本结构与工作原理，了解纳米材料在化学电容器中的储能机理，熟悉化学电容器的常用表征技术。

（5）熟悉燃料电池的基本结构和工作原理，了解纳米材料在燃料电池中的电催化原理，熟悉电催化材料的常用电化学表征技术。

（6）能够从宏观和微观的不同角度理解化学变化的基本特征，培养逻辑思维能力和批判性思维精神。

（7）培养学生理论联系实际的能力和实事求是的科学态度、终身学习意识和自我管理、自主学习能力。

（8）培养学生的实验动手能力、发现问题和解决问题的能力。

（1）Be acquainted with the basic methods, principles, and characteristics of nanomaterial synthesis, understand the relationship between the composition, structure, and energy storage properties of nanomaterials, and apply these theories to explain the mechanism of nanosynthesis, and the relationship between structure and chemical energy storage properties, and their variation patterns.

（2）Master the common synthesis methods and basic principles of nanomaterials, and grasp the energy storage principles of nanomaterials in energy storage devices.

（3）Learn the basic structure and working mechanism of lithium-ion batteries, and grasp the preparations of cathodic and anodic materials, and the common characterization techniques.

（4）Learn the basic structure and working mechanism of chemical capacitors, the energy storage mechanism of nanomaterials, and the common characterization techniques.

（5）Learn the basic structure and working mechanism of fuel cells, electrocatalytic mechanism of nanomaterials, and the common electrochemical characterization techniques.

（6）Understand the basic mechanism of chemical reactions from macro and micro perspectives and develop logical thinking ability and critical thinking spirit.

（7）Develop the ability to link practice and theory, and to cultivate scientific attitude of "seeking truth from facts", lifelong learning awareness, self-management, and autonomous learning.

（8）Develop the ability to conduct experiments, discovering, and solving problems.

6.6.4　课程大纲和知识点（Syllabus and Key Points）

绪论（introduction）

章节序号 chapter number	章节名称 chapters	知识点 key points	课时 class hour
0	绪论 introduction	（1）纳米科学技术的发展史 （2）纳米材料的特征 （3）纳米材料的分类 （4）化学储能概述 （1）history of nanoscience and nanotechnology （2）characteristics of nanomaterials （3）classification of nanomaterials （4）overview of chemical energy storage	2

第一章 纳米材料的结构、性质及表征（structure, properties, and characterization of nanomaterials）

章节序号 chapter number	章节名称 chapters	知识点 key points	课时 class hour
1.1	纳米材料的结构、性质及表征 structure, properties, and characterization of nanomaterials	（1）纳米材料的结构特点 （2）纳米材料的基本性质 （3）纳米材料的表征 （1）structural characteristics of nanomaterials （2）basic properties of nanomaterials （3）characterization of nanomaterials	2

第二章 纳米材料的气相制备方法（gas-phase preparation methods of nanomaterials）

章节序号 chapter number	章节名称 chapters	知识点 key points	课时 class hour
2.1	纳米材料气相制备方法 gas-phase preparation methods of nanomaterials	（1）物理气相法 （2）化学气相反应法 （1）physical vapor method （2）chemical vapor reaction method	2

第三章 纳米材料的液相制备方法（liquid-phase preparation methods of nanomaterials）

章节序号 chapter number	章节名称 chapters	知识点 key points	课时 class hour
3.1	纳米材料液相制备方法 liquid-phase preparation methods of nanomaterials	（1）溶胶-凝胶法 （2）冷冻干燥法 （3）喷雾法 （4）溶液沉淀法 （5）水热-溶剂热法 （6）静电纺丝法 （1）sol-gel method （2）freeze-drying method （3）spray method （4）solution precipitation method （5）hydrothermal-solvothermal method （6）electrospinning method	4

第四章　纳米材料的固相制备方法（solid-phase preparation methods of nanomaterials）

章节序号 chapter number	章节名称 chapters	知识点 key points	课时 class hour
4.1	纳米材料固相制备方法 solid-phase preparation methods of nanomaterials	（1）粉碎法 （2）热分解法 （3）固相反应法 （4）其他方法 （1）crushing method （2）thermal decomposition method （3）solid-phase reaction method （4）other methods	2

第五章　纳米材料在锂离子电池中的应用（application of nanomaterials in lithium-ion batteries）

章节序号 chapter number	章节名称 chapters	知识点 key points	课时 class hour
5.1	纳米材料在锂离子电池 中的应用 application of nanomaterials in lithium-ion batteries	（1）锂离子电池简介 （2）锂离子电池的工作原理 （3）锂离子电池用负极材料 （4）锂离子电池用正极材料 （5）锂离子电池常用表征技术 （1）introduction to lithium-ion batteries （2）working principle of lithium-ion batteries （3）anode materials for lithium-ion batteries （4）cathode materials for lithium-ion batteries （5）commonly used characterization techniques for lithium-ion batteries	4

第六章　纳米材料在化学电容器中的应用（application of nanomaterials in chemical capacitors）

章节序号 chapter number	章节名称 chapters	知识点 key points	课时 class hour
6.1	纳米材料在化学电容器 中的应用 application of nanomaterials in chemical capacitors	（1）化学电容器简介与分类 （2）化学电容器的工作原理 （3）化学电容器常用电极材料 （4）化学电容器的表征技术	4

续表

章节序号 chapter number	章节名称 chapters	知识点 key points	课时 class hour
6.1	纳米材料在化学电容器中的应用 application of nanomaterials in chemical capacitors	（1）introduction and classification of chemical capacitors （2）working principle of chemical capacitors （3）electrode materials commonly used in chemical capacitors （4）characterization technology of chemical capacitors	4

第七章 纳米材料在燃料电池中的应用（application of nanomaterials in fuel cells）

章节序号 chapter number	章节名称 chapters	知识点 key points	课时 class hour
7.1	纳米材料在燃料电池中的应用 application of nanomaterials in fuel cells	（1）燃料电池简介 （2）燃料电池的基本工作原理 （3）燃料电池用电催化剂材料 （4）燃料电池用电催化剂的表征技术 （1）introduction of fuel cells （2）basic working principle of fuel cells （3）electrocatalyst materials for fuel cells （4）characterization technology of electrocatalyst for fuel cells	4

6.6.5 实验环节（Experiments）

序号 number	实验内容 experiment content	知识点 key points	课时 class hour
1	无定型碳纳米材料的合成与表征 synthesis and characterization of amorphous carbon nanomaterials	（1）前驱体合成 （2）前驱体热处理制备碳纳米材料 （3）结构与组成表征 （1）precursor synthesis （2）precursor heat treatment to prepare carbon nanomaterials （3）characterization of structure and composition	4

序号 number	实验内容 experiment content	知识点 key points	课时 class hour
2	碳基纳米材料的锂离子电池性能研究 study on the performance of carbon-based nanomaterials for lithium-ion batteries	（1）电极浆料制备 （2）工作电极制备 （3）锂离子电池封装 （4）电池性能评价 （1）preparation of electrode slurry （2）preparation of working electrode （3）lithium-ion battery package （4）battery performance evaluation	4
3	碳基纳米材料的超级电容器性能研究 study on the performance of carbon-based nanomaterial for supercapacitors	（1）电极浆料制备 （2）工作电极制备 （3）超级电容器性能评价 （1）preparation of electrode slurry （2）preparation of working electrode （3）supercapacitor performance evaluation	4
4	碳基纳米材料的氧还原性质研究 study on oxygen reduction properties of carbon-based nanomaterials	（1）电极浆料制备 （2）工作电极制备 （3）氧还原性能评价 （1）preparation of electrode slurry （2）preparation of working electrode （3）oxygen reduction performance evaluation	4

6.7 "固态电池"教学大纲

课程名称：固态电池

Course Title：Solid State Batteries

先修课程：电化学基础

Prerequisites：Fundamentals of Electrochemistry

学分：2

Credits：2

6.7.1 课程目的和基本内容（Course Objectives and Basic Contents）

本课程是储能科学与工程本科专业的一门选修课程。课程以固体物质的晶体学性质、电子电导、离子输运为基础探讨固态电池的前沿科学问题。

This course is an elective course for the undergraduate majoring in energy storage science and engineering.The course focuses on the frontier scientific issues of solid-state batteries based on the crystallographic properties, electronic conductance, and ion transport of solid state materials.

6.7.2 课程基本情况（ Basic Information of the Course ）

课程名称	固态电池 Solid State Batteries							
开课时间	一年级	二年级	三年级	四年级	总学分	2	总学时	40
	秋 春	秋 春	秋 春	秋 春				
课程定位	储能科学与工程专业本科生电化学与电磁储能课程群（B模块）选修课							
授课学时分配	课堂讲授32学时+实验8学时							
先修课程	电化学基础							
后续课程	无							
教学方式	课堂教学，大作业，实验与综述报告，讨论，平时作业							
考核方式	课程结束笔试成绩占50%，实验报告占10%，平时成绩占40%							
参考教材	[1] 杨勇.固态电化学.北京:化学工业出版社,2016. [2] 崔光磊.动力锂电池中聚合物关键材料.北京:科学出版社,2018.							
参考资料	[1] 吴玉平,袁翔云,董超,等.锂离子电池:应用与实践.北京:化学工业出版社,2011. [2] 张剑波,李哲,吴彬.锂离子电池结构设计理论与应用.北京:中国科学技术出版社,2016. [3] Masaki Yoshio, Brodd R J, Akiya Kozawa. Lithium-Ion Batteries: Science and Technologies. Berlin: Springer, 2009. [4] Tsutomu Minami, Masahiro Tatsumisago, Masataka Wakihara, et al. Solid State Ionics for Batteries. Tokyo: Springer, 2005. [5] Dudney N J,West W C, Nanda J. Handbook of Solid State Batteries. Singapore: World Scientific Publishing Co. Pte. Ltd., 2015. [6] Julien C, Nazri G A. Solid State Batteries: Materials Design and Optimization. Berlin: Springer, 1994.							

6.7.3 教学目的和基本要求（ Teaching Objectives and Basic Requirements ）

通过对本课程的学习，使学生了解固态电池原理、固态电池的材料体系、固态电池的组装与测试方法、固态电池的表征技术以及固态电池的优点与存在问题；通过小组作业及实验，培养学生自学和归纳总结能力、运用所学知识分析问题和解决问题的能力以及团队合作精神；培养学生理论联系实际的能力和实事求是的科学态度、终身学习意识和自我管理、自主学习能力，培养学生逻辑思维能力和批判性思维精神。

基本要求：

（1）掌握固态电解质传导离子机理，了解固态电解质的构效关系；掌握固态电解质的种类以及制备和开发方法。

（2）掌握固态电解质与电极界面的研究方法，掌握氧化物 / 硫化物等无机固态电解质结构及电化学性质。

（3）掌握聚合物固态电解质结构及电化学性质，掌握含锂型过渡金属氧化物及无锂型电极活性材料的结构和锂离子脱嵌性质。

（4）掌握固态材料中电子 / 离子传导的基本概念，能区分电子 / 离子电导；掌握固态脱嵌反应的机理及动力学问题。

（5）掌握固态体系中电化学的基本概念、原理和应用；掌握固 – 固界面离子输运的基本知识，能够从宏观和微观的不同角度理解界面传导、离子脱嵌的基本特征。

This course enables students to understand the principle of solid state batteries, the material system of solid state batteries, the assembly and testing methods of solid state battery, the characterization technology of solid state batteries, and the advantages and problems of solid state batteries. Through group assignments and experimental questions, the students can develop self-learning, inductive summarization, teamwork skills, and obtain the ability to employ what they have learned to analyze and solve problems. The course cultivates students' ability to integrate theory with practice, develop a scientific attitude to seek truth from facts, develop lifelong learning consciousness and self-management skills. It also cultivates students' independent learning altitude, and logical thinking ability and critical thinking spirit.

Basic requirements:

（1）Master the ion conduction mechanism of solid electrolytes, understand the structure-activity relationship of solid electrolytes; master the types and preparation methods of solid electrolytes.

（2）Master the research methods of solid electrolyte and electrode interface; master the structure and electrochemical properties of inorganic solid electrolytes such as oxides/sulfides.

（3）Master the structure and electrochemical properties of polymer solid electrolytes; master the structure and lithium-ion deintercalation properties of lithium-containing transition metal oxides and lithium-free electrode active materials.

（4）Master the basic concepts of electronic/ionic conduction in solid materials, and distinguish electronic/ionic conductance; master the mechanism and kinetics of solid deintercalation reactions; master the basic concepts, principles, and applications of electrochemistry in solid system.

（5）Master the basic concepts, principle, and applications of electrochemistry in solid system; mater the basic knowledge of ion transport at the solid/solid interface, and understand the basic characteristics of interface conduction and ion deintercalation from macro and micro perspectives.

6.7.4 课程大纲和知识点（Syllabus and Key Points）

第一章 绪论（introduction）

章节序号 chapter number	章节名称 chapters	知识点 key points	课时 class hour
1.1	课程规划,固态电池的兴起 course planning, rise of solid-state batteries	固态电池的发展历史及应用 development history and application of solid-state batteries	1

第二章 固态电池中的科学（science in solid-state batteries）

章节序号 chapter number	章节名称 chapters	知识点 key points	课时 class hour
2.1	固态电池概述 overview of solid-state batteries	（1）固态电池基本结构 （2）脱嵌反应动力学 （3）能带的概念 （4）离子的概念 （5）离子电导的理论 （6）扩散的概念及理论模型 （7）扩散的类型及特点 （8）固态离子扩散特性与影响因素 （9）电子电导与离子电导的区分 （1）basic structure of solid-state batteries （2）dynamics of deintercalation reaction （3）concept of energy bands （4）concept of ions （5）theory of ionic conductance （6）concept and theoretical model of diffusion （7）types and features of diffusion （8）characteristics and influencing factors of solid ion diffusion （9）difference between electronic conductance and ionic conductance	5

第三章　固态电解质与界面（solid electrolytes and interfaces）

章节序号 chapter number	章节名称 chapters	知识点 key points	课时 class hour
3.1	固态电解质体系 solid electrolyte system	（1）无机固态电解质 （2）聚合物固态电解质 （3）复合固态电解质 （4）固体电解质制备方法与技术 （5）当前研究热点问题探讨 （1）inorganic solid electrolyte （2）polymer solid electrolyte （3）composite solid electrolyte （4）preparation method for solid electrolyte （5）discussion for current research hot issues	6
3.2	固态电池中的界面 interfaces in solid state batteries	（1）固-固界面类型 （2）固体电极-固体电解质界面 （3）界面电荷传输分析方法 （4）电极-固体电解质界面阻抗 （5）电荷转移阻抗测定 （6）减小电荷转移阻抗方法 （7）活化能 （8）可控界面构建方法 （9）当前研究进展和发展趋势 （1）type of solid/solid interface （2）electrode/solid electrolyte interface （3）characterization methods for interfacial charge transport （4）electrode/solid electrolyte interfacical impedance （5）measurement for the charge-transfer resistance （6）approaches for reducing the charge-transfer resistance （7）activation energy （8）controllable interface construction method （9）currently related research progress and development tendency	6

第四章　固态电池的组装与测试（assembly and measurement of solid state batteries）

章节序号 chapter number	章节名称 chapters	知识点 key points	课时 class hour
4.1	固态电池的组装 assembly of solid state batteries	（1）三维电池 （2）全固态薄膜电池 （3）结构电池	4

续表

章节序号 chapter number	章节名称 chapters	知识点 key points	课时 class hour
4.1	固态电池的组装 assembly of solid state batteries	（4）银固态电池 （1）three-dimensional batteries （2）all solid state thin-film batteries （3）structural batteries （4）silver solid state batteries	4
4.2	固态电池的测试 measurements of solid state batteries	（1）离子和电子电导率测试 （2）迁移数确定方法 （3）热力学和动力学测量 （1）measurement for ionic electronic conductivity （2）determining methods for transference number （3）measurements for thermodynamic and kinetic	4

第五章 固态电池中的表征技术（characterisation techniques for solid state batteries）

章节序号 chapter number	章节名称 chapters	知识点 key points	课时 class hour
5.1	形貌与表面分析 morphology and surface analysis	（1）扫描电镜 （2）透射电镜 （1）scanning transmission electron microscopy （2）transmission electron microscopy	2
5.2	结构分析 structure analysis	（1）固体核磁 （2）中子衍射 （3）同步辐射X射线 （4）三维成像 （1）solid state NMR （2）neutron diffraction （3）synchrotron X-ray diffraction （4）three-dimensional imaging	2
5.3	理论计算模拟 theoretical computational simulation	（1）材料模拟计算的理论基础 （2）常见的模拟方法 （3）第一性原理计算模拟 （1）theoretical basis of material simulation calculation （2）common simulation methods （3）first-principles computational simulation	2

6.7.5　实验环节（Experiments）

序号 number	实验内容 experiment content	知识点 key points	课时 class hour
1	材料合成与测试 materials synthesis and characterization	（1）电极材料的制备方法，物相鉴定，形貌观测，扫描电镜，X射线衍射 （2）固态电解质及电极材料晶体结构及微观结构 （1）preparation method of electrode materials, phase identification, morphological observation, scanning electron microscope, X-ray diffraction （2）crystal structure and micro structure of solid electrolyte and electrode materials	4
2	固态电池组装 solid state batteries assembly	（1）固态电池组装方法，电化学测试 （2）材料结构与性能的关系 （1）solid state batteries assembly method, electrochemical test （2）relationship between material structure and performance	4

_ 第 7 章 _

系统储能课程群

7.1 "储能系统设计"教学大纲

课程名称：储能系统设计

Course Title：Design of Energy Storage System

先修课程：电路，现代电子技术，储能原理Ⅰ，储能原理Ⅱ

Prerequisites：Electric Circuit, Modern Electronic Technology, Energy Storage PrincipleⅠ, Energy Storage PrincipleⅡ

学分：2

Credits：2

7.1.1 课程目的和基本内容（Course Objectives and Basic Contents）

本课程是储能科学与工程专业的一门专业必修课程。课程以先进电池管理系统开发为背景，涵盖储能系统发展、电池储能系统的组成原理、电池荷电状态估算、储能系统均衡模式设计等内容。其任务是使学生了解储能系统发展，掌握电池储能系统组成及设计的基础知识和技能，进而可以独立设计包括均衡、检测等功能的储能电池管理系统，为学生的学习、工作和进一步提高打下坚实的基础。

This course is a mandatory course for undergraduates majoring in Energy Storage Science and Engineering.The course focuses on the development of advanced battery management systems. It covers the development of energy storage system (ESS), the principle of battery energy storage system, state of charge (SoC) estimation, design of equilibrium mode etc. The objective of this course is to train students to understand the development of energy storage system, master the fundamental knowledge and skills on design of battery energy storage system (BESS) and battery management system (BMS) independently, which will be beneficial for further study and work after graduation.

7.1.2 课程基本情况（Basic Information of the Course）

课程名称	储能系统设计 Design of Energy Storage System								
开课时间	一年级		二年级		三年级		四年级		总学分	2	总学时	40
	秋	春	秋	春	秋	春	秋	春			
课程定位	储能科学与工程专业本科生储能系统课程群必修课								
授课学时分配	课堂讲授32学时+实验8学时								
先修课程	电路,现代电子技术,储能原理Ⅰ,储能原理Ⅱ								
后续课程	储能装置设计与开发实验								
教学方式	课堂教学,实验,讨论								
考核方式	课程结束笔试成绩占60%,实验报告占20%,平时成绩占20%								
参考教材	[1] 瑞恩,王朝阳.电池建模与电池管理系统设计.北京:机械工业出版社,2018.								
参考资料	[1] 吴福保,杨波,叶季蕾.电力系统储能应用技术.北京:中国水利水电出版社,2014. [2] 王兆安,刘进军.电力电子技术.北京:机械工业出版社,2009. [3] 蔡旭,李睿,李征.储能功率变换与并网技术.北京:科学出版社,2019.								

7.1.3 教学目的和基本要求（Teaching Objectives and Basic Requirements）

通过本课程的学习，使学生获得储能系统设计的基本理论和知识，为后续相关课程的学习打下坚实的基础。

基本要求：

（1）掌握储能系统设计的概念和内容；

（2）了解储能系统发展、电池储能系统的组成原理；

（3）掌握电池荷电状态估算的方法；

（4）掌握储能系统均衡模式设计方法；

（5）掌握储能电池管理系统设计方法；

（6）掌握电池储能功率转换系统设计方法。

Through the study of this course, students can obtain the basic theory and knowledge of ESS design, and lay a solid foundation for the subsequent related courses.

Basic requirements:

（1）Master the concept and content of ESS design.

（2）Understand the development of ESS and the principle of BESS.

（3）Master the estimation method of SoC.

（4）Master the equilibrium design method of BESS.

（5）Master the design method of BMS.

（6）Master the design method of power conversion system on BESS.

7.1.4 课程大纲和知识点（Syllabus and Key Points）

第一章 绪论（introduction）

章节序号 chapter number	章节名称 chapters	知识点 key points	课时 class hour
1.1	储能系统的发展 development of energy storage system (ESS)	（1）储能在电力系统中的应用 （2）主要的电力储能技术 （1）application of energy storage in power system （2）overview of power storage technology	1
1.2	电池储能系统发展 development of battery energy storage system (BESS)	（1）电池的类型与作用 （2）电池储能系统工程 （1）battery type and role （2）BESS engineering	1

第二章 电池储能系统的组成原理（principle of battery energy storage system）

章节序号 chapter number	章节名称 chapters	知识点 key points	课时 class hour
2.1	电池储能系统的基本结构 basic structure of battery energy storage system (BESS)	电池储能系统的基本结构 basic structure of BESS	1
2.2	电池管理系统的功能与组成 function and composition of battery management system (BMS)	（1）电池管理系统的功能 （2）电池管理系统的组成 （1）function of BMS （2）composition of BMS	1
2.3	储能功率变换系统 power conversion system (PCS)	（1）储能功率变换系统的功能 （2）储能功率变换系统的基本结构 （1）function of PCS （2）structure of PCS	2

第三章　电池荷电状态估算（estimation of battery state of charge）

章节序号 chapter number	章节名称 chapters	知识点 key points	课时 class hour
3.1	电池荷电状态（SoC）估计 estimation of battery state of charge (SoC)	（1）电池的一般电路模型 （2）SoC的电流积分法 （3）SoC的电压查表法 （1）circuit model of battery （2）SoC current integral method （3）SoC voltage look-up table method	2
3.2	常见电池的SoC估算方法 SoC estimation method of conventional batteries	（1）SoC计算的影响因素 （2）锂离子电池的SoC估算 （3）铅酸电池的SoC估算 （1）influencing factors on SoC calculation （2）SoC estimation of lithium-ion batteries （3）SoC estimation of lead acid batteries	2

第四章　储能系统均衡模式的设计（design of equilibrium mode for energy storage system）

章节序号 chapter number	章节名称 chapters	知识点 key points	课时 class hour
4.1	主要均衡模式 main equilibrium modes	（1）串联均衡模式 （2）并联均衡模式 （1）series equilibrium mode （2）parallel equilibrium mode	4
4.2	储能电池均衡管理系统的设计 equilibrium management system design for energy storage battery	（1）均衡管理系统的需求 （2）均衡管理系统的方案设计 （3）均衡模块设计及连接方式 （1）requirements of equilibrium management system （2）design for equilibrium management system （3）connections and design of equilibrium modules	4

第五章 储能电池检测与管理系统设计（design of energy storage battery detection and management system）

章节序号 chapter number	章节名称 chapters	知识点 key points	课时 class hour
5.1	电池组的参量检测 parameters detection of battery pack	（1）电池组的主要参量 （2）电池组参量常用检测方法 （1）parameters of battery pack （2）conventional detection methods for battery pack parameters	4
5.2	SoH估计 estimation of state of health (SoH)	（1）电池SoH的主要估算方法 （2）基于阻抗的估算方法及实现 （1）main SoH estimation methods （2）estimation method and realization based on impedance	2
5.3	电池管理系统设计 design of battery management system (BMS)	（1）电池管理系统的总体结构 （2）电池管理系统的硬件设计 （3）电池管理系统的软件设计 （1）overall structure of BMS （2）hardware design of BMS （3）software design of BMS	4

第六章 储能电池功率转换系统（power conversion system for battery energy storage）

章节序号 chapter number	章节名称 chapters	知识点 key points	课时 class hour
6.1	PCS的典型电路结构 typical circuit structure of power conversion system(PCS)	（1）基于△/Y变压器拓扑 （2）三单相变压器组合式拓扑 （3）基于直流母线分裂电容的主电路拓扑 （4）基于级联H桥的中高压拓扑 （1）△/Y transformer topology （2）three-phase transformer combined topology （3）circuit topology based on DC bus split capacitor （4）medium and high voltage topology based on cascaded H-bridge	2
6.2	PCS控制技术 control technology of PCS	（1）PCS控制技术概述 （2）V/f控制 （3）PQ控制 （4）下垂控制 （1）overview of PCS control technology （2）V/f control （3）PQ control （4）droop control	2

7.1.5　实验环节（Experiments）

序号 number	实验内容 experiment content	知识点 key points	课时 class hour
1	电池串联均衡实验 equalization experiment for series batteries	（1）设定串联电池不同SoC状态，完成串联均衡电路设计 （2）对比不同均衡电路参数对电池均衡效果的影响 （1）completing equali zation circuit design for batteries in series based on different SoC states （2）investigating the battery equalization performance of different equalization circuit parameters	4
2	电池储能系统实验 experiment for battery energy storage system (BESS)	（1）了解锂离子电池、铅酸电池的充放电原理，分析充放电的影响因素 （2）学习实验平台不同充电模式的设定和操作方法，对电池进行充放电实验 （3）总结不同充电电流对电池性能的影响 （1）knowing the charging and discharging principles of lithium-ion battery and lead-acid battery；analyzing different influencing factors on charging and discharging processes （2）grasping the setting and operation methods for different charging modes, and completing charging and discharging experiments on batteries （3）summarizing the effect of different charging currents on battery performance	4

7.2　"电力系统分析"教学大纲

课程名称：电力系统分析
Course Title：Power System Analysis
先修课程：高等数学Ⅰ，大学物理（含实验），电路
Prerequisites：Advanced Mathematics Ⅰ, Physics and Physics Experiments,Electric Circuit
学分：5
Credits：5

7.2.1　课程目的和基本内容（Course Objectives and Basic Contents）

本课程的目的是使学生掌握电力系统的基本理论和计算方法，培养学生综合分析和解决电力系统工程问题的能力，为日后从事电力系统的研究和工程应用打下基础。

课程的主要内容包括电力系统的基本概念和基础理论知识，电网的元件参数、等值电路、运行特性和计算方法，电力系统的潮流计算，电力系统正常运行方式的调整与控制。

The objective of this course is to let students master the knowledge of theoretical fundamentals and skills of computation techniques for power systems, and develop the capability of comprehensive analysis and solving engineering problems in power systems, which build the basis for students to conduct research and engineering applications in the future.

The content of the course includes the concept and fundamentals of power systems, power grid component parameters, equivalent circuit, operation features and associated computation methods, power flow computation, and adjustment and control of power system operation under normal conditions.

7.2.2 课程基本情况（Basic Information of the Course）

课程名称	电力系统分析 Power System Analysis									
开课时间	一年级		二年级		三年级		四年级		总学分	5	总学时	84
	秋	春	秋	春	秋	春	秋	春				
课程定位	储能科学与工程专业本科生储能系统课程群必修课											
授课学时分配	课堂讲授80学时+上机4学时											
先修课程	高等数学Ⅰ,大学物理（含实验）,电路											
后续课程	智能电网储能应用技术,能源互联网											
教学方式	课堂教学,平时作业,上机编程											
考核方式	课程结束笔试成绩占70%,平时成绩占30%											
参考教材	[1]夏道止,杜正春.电力系统分析.3版.北京:中国电力出版社,2018.											
参考资料	[1]陈珩.电力系统稳态分析.4版.北京:中国电力出版社,2018. [2]陆敏政.电力系统习题集.北京:水利电力出版社,1990. [3]何仰赞,温增银.电力系统分析:上、下册.4版.武汉:华中科技大学出版社,2018. [4]韩祯祥.电力系统分析.5版.杭州:浙江大学出版社,2013. [5]冯慈璋,马西奎.工程电磁场导论.北京:高等教育出版社,2000. [6]阎治安,苏少平,崔新艺.电机学.西安:西安交通大学出版社,2018. [7] Stevenson W D. Elements of Power System Analysis. New York: McGraw-Hill Companies,Inc., 1975. [8] Glover J D, Sarma M S, Overbye T. Power system analysis and design.Boston:Cengage Learning, 2012.											

7.2.3　教学目的和基本要求（Teaching Objectives and Basic Requirements）

（1）全面了解电力系统中各元件，对各元件的特性、作用、相互间关系，用全局的观点去认识和了解，学会综合地分析电力系统工程问题的基本方法；

（2）巩固、理解先修课程有关的理论及方法，并掌握这些理论和方法在电力系统分析课程中的应用；

（3）训练工程计算能力，培养学生分析和解决电力系统工程问题的能力；

（4）掌握运用电子计算机解决电力系统工程问题的理论和方法；

（5）了解电力系统的发展趋势及电力系统中的新技术，提高学生自主学习的意识。

（1）Systematically master the knowledge of components in power systems, understand their features, effects, and relationship in a systematical perspective, and cultivate the skills of analyzing engineering problems of power systems comprehensively.

（2）Review the theory and methods of prerequisites, and understand how they are applied in the power system analysis.

（3）Train the students' capabilities of engineering computation, analyzing and solving engineering problems in power systems.

（4）Master the capability of solving power system engineering problems using computers.

（5）Understand the trend in power system development and new techniques and enhance their awareness of self-directed learning.

7.2.4　课程大纲和知识点（Syllabus and Key Points）

第一章　电力系统的基本概念（concept of power systems）

章节序号 chapter number	章节名称 chapters	知识点 key points	课时 class hour
1.1	电力系统简介、电力系统构成 introduction to power systems and their structure and components	（1）电力系统简介 （2）电力系统的构成 （1）introduction to power systems （2）structure and components of power systems	2
1.2	电力系统的接线方式及电压等级 wiring methods and voltage levels of power systems	（1）电力系统的接线方式 （2）电力系统的电压等级 （1）wiring methods in power systems （2）voltage levels of power systems	2

续表

章节序号 chapter number	章节名称 chapters	知识点 key points	课时 class hour
1.3	电力系统的特点和运行的基本要求 features of power systems and the basic requirements of their operation	（1）发电厂生产过程 （2）电力系统负荷 （3）电力系统运行的基本要求 （1）producing process of power plants （2）load of power systems （3）basic requirements of power system operation	2
1.4	我国的电力系统和电力市场简介 introduction to the domestic power systems and power market	（1）我国电力系统概况 （2）电力市场的概念和研究问题 （1）introduction to domestic power systems （2）concept of the electricity market and related research problems	2

第二章 电力系统基础理论知识（fundamental knowledge of power systems）

章节序号 chapter number	章节名称 chapters	知识点 key points	课时 class hour
2.1	电场基本知识 fundamentals of electric field	（1）电场强度,高斯定律,电场基本方程 （2）电容的概念和计算 （1）electric filed intensity, Gauss' law, basic equations of electric field （2）concept and computation of capacitors	4
2.2	磁场基本知识 fundamentals of magnetic field	（1）磁感应强度,安培环路定律,磁场基本方程 （2）电感的概念和计算 （1）magnetic intensity, Ampere's circuital law, basic equations of magnetic field （2）concept and computation of inductors	4
2.3	变压器原理及基本理论 mechanism and theoretical fundamentals of transformers	（1）变压器的结构和工作原理 （2）变压器的基本理论 （1）structures of transformers and their working mechanism （2）theoretical fundamentals of transformers	4
2.4	发电机原理及基本理论 mechanism and theoretical fundamentals of generators	（1）发电机的结构和工作原理 （2）发电机的基本理论 （1）structure of generators and their working mechanism （2）theoretical fundamentals of generators	4

第三章　电力网的正序参数和等值电路（positive sequence parameters and equivalent circuit of power grid）

章节序号 chapter number	章节名称 chapters	知识点 key points	课时 class hour
3.1	电力线路参数、输电线路模型及等值电路 parameters of transmission lines, models of transmission lines, and their equivalent circuit	（1）电力线路参数：电阻、电抗、电纳、电导 （2）输电线路方程及等值电路 （1）parameters of transmission lines: resistor, reactance, susceptance, conductance （2）equations of transmission lines and their equivalent circuit	6
3.2	变压器参数及等值电路 parameters of transformers and equivalent circuit	（1）变压器参数 （2）变压器等值电路 （1）parameters of transformers （2）equivalent circuit of transformers	6
3.3	多电压等级电力网的等值电路及标幺制 per-unit equivalent circuit of power grid with multiple voltage levels and per-unit scheme	（1）标幺制 （2）多电压等级电网标幺制等值电路 （1）per-unit scheme （2）per-unit equivalent circuit of power grid with multiple voltage levels	4

第四章　输电线路运行特性及简单电力网络潮流估算（transmission grid operation characteristics and approximate computation method for power flow）

章节序号 chapter number	章节名称 chapters	知识点 key points	课时 class hour
4.1	电力网的功率损耗和电压降落 power loss and voltage drops of power grid	（1）电压降落和损耗 （2）功率分布和功率损耗 （1）voltage drop and voltage loss （2）distribution of power and power loss	4
4.2	输电线路的运行特性 operation characteristics of transmission lines	（1）输电线路空载运行特性 （2）输电线路传输功率极限 （1）no-load operation characteristics of transmission lines （2）power transmission limit	3
4.3	简单辐射性网络和闭式网络的潮流估算方法 simple approximate computation method of power flow for radial and looped network	（1）辐射性网络的潮流估算方法 （2）闭式网络的潮流估算方法 （1）approximate computation method of power flow for network with radial topology （2）approximate computation method of power flow for network with looped topology	3

第五章 电力系统潮流的计算机算法（computer-based algorithms for power flow computation）

章节序号 chapter number	章节名称 chapters	知识点 key points	课时 class hour
5.1	电力网的数学模型,节点导纳矩阵,节点阻抗矩阵 mathematical models for power grid, nodal admittance matrix, nodal impedance matrix	（1）用节点导纳矩阵表示的网络方程 （2）用节点阻抗矩阵表示的网络方程 （1）power network equation using nodal admittance matrix （2）power network equation using nodal impedance matrix	3
5.2	电力系统潮流计算中功率方程和节点分类 power balance equation and category of nodes in power flow computation	（1）直角坐标和极坐标下的节点功率方程 （2）潮流计算中节点分类 （1）nodal power balance equation under rectangular coordinates and polar coordinates （2）category of nodes in power flow computation	4
5.3	潮流计算的牛顿-拉弗森法 Newton-Raphson method for power flow computation	（1）牛顿-拉弗森算法的基本原理 （2）牛顿-拉弗森潮流计算算法的步骤 （1）principles of Newton-Raphson method （2）process of power flow computation using Newton-Raphson method	4
5.4	牛顿-拉弗森法潮流计算的收敛性和稀疏技术 convergence of Newton-Raphson method and sparsity technique	（1）初值、收敛性和多值解 （2）稀疏矩阵技术 （1）initial value, convergence, and multiple solutions （2）sparse matrix technique	4
5.5	其他潮流计算方法简介 brief introduction of other power flow computation methods	（1）潮流计算快速解耦法 （2）直流潮流 （1）fast decoupling method for power flow computation （2）direct current (DC) power flow	3

第六章 电力系统正常运行方式的调整和控制（adjustment and control of power system operation under normal condition）

章节序号 chapter number	章节名称 chapters	知识点 key points	课时 class hour
6.1	电力系统有功功率和频率的调整和控制 control of real power and frequency in power systems	（1）有功功率和频率之间关系 （2）频率的一次调整 （3）频率的二次、三次调整 （1）relation between real power and frequency （2）primary frequency control （3）secondary and tertiary frequency control	4

章节序号 chapter number	章节名称 chapters	知识点 key points	课时 class hour
6.2	电力系统无功功率和电压的调整和控制 reactive power and adjustment and control method of voltage in power systems	（1）负荷的静态电压特性 （2）无功补偿设备特性 （3）电压调整和控制方法 （1）static voltage features of load （2）features of reactive power compensation （3）adjustment and control method of voltage in power systems	4
6.3	电力系统运行方式的优化 optimization for power system operation	（1）有功功率经济分配 （2）最优潮流 （1）economic dispatch of real power （2）optimal power flow	4
6.4	电力系统潮流控制和高压直流输电 power flow control and high voltage direct current transmission system	（1）高压直流输电 （2）柔性交流输电 （3）FACTS设备的原理 （1）high voltage direct current (HVDC) transmission system （2）flexible alternate current transmission system (FACTS) （3）principles of FACTS devices	4

7.2.5　上机实验环节（Experiments on Computers）

序号 number	实验内容 experiment content	知识点 key points	课时 class hour
1	潮流计算的上机实验 experiments on computer for power flow computation	（1）牛顿-拉弗森法和PQ分解法的步骤 （2）编程实现两种算法 （1）steps of Newton-Raphson method and PQ decomposition algorithm for power flow computation （2）implementation of the two algorithms on computers	4

7.3　"储能系统检测与估计"教学大纲

课程名称：储能系统检测与估计

Course Title：Detection and Estimation of Energy Storage Systems

先修课程：高等数学Ⅰ，线性代数与解析几何，概率论与数理统计，自动控制理论，储能原理Ⅰ，储能原理Ⅱ

Prerequisites：Advanced Mathematics Ⅰ, Linear algebra and analytical geometry, Probability Theory and Mathematical Statistics, Principles of Automatic Control，Energy Storage Principle Ⅰ, Energy Storage Principle Ⅱ

学分： 2

Credits： 2

7.3.1 课程目的和基本内容（Course Objectives and Basic Contents）

本课程是储能科学与工程专业的一门专业基础必修课。本课程的任务是使学生掌握储能系统检测与估计领域基础知识和理解能力，使学生具备储能系统检测与估计工程问题分析能力，并能够利用专业知识结合文献研究，解决复杂工程问题。

基本要求：

（1）储能系统检测与估计领域基础知识理解能力。帮助学生掌握储能系统检测领域的基础知识，包括储能检测技术的理论基础与储能系统传感器的技术基础；掌握储能系统估计的理论基础，包括储能系统估计理论的数学基础与系统估计的理论基础；熟练掌握储能系统检测与估计的方差与均方误差的计算方法；并熟练掌握线性动态系统估计的方法与非线性滤波点的估计方法。

（2）具备储能系统检测与估计工程问题分析能力。能够运用储能系统检测与估计领域的相关知识，对储能系统检测与估计在实际应用中的具体工程问题进行分析。能够在实际工程问题中构建储能系统检测与估计的线性动态系统估计模型与非线性滤波点估计模型，并根据实际的模型特点提出有效的故障检测、辨识、估计与状态估计方案。

（3）利用专业知识结合文献研究，解决复杂工程问题的能力。针对储能系统检测与估计领域的复杂工程问题，能够利用储能系统检测与估计领域的基本知识，通过文献检索、调研等科学方法进行研究，合理选择储能系统检测与估计的模型体系，制订研究路线，设计实验方案，安全有效地开展实验，针对储能系统检测与估计问题设计合理有效的故障检测、辨识、估计与状态估计方案。

This course is a mandatory fundamental course for undergraduates majoring in Energy Storage Science and Engineering. The course enables undergraduates to master the basics of detection and estimation for energy storage systems, so that students are capable of analyzing the engineering problems of energy storage system detection and estimation, and using their expertise and literature studies to solve complex engineering problems.

Basic requirements:

（1）Ability to understand basic knowledge in the field of energy storage system detection and estimation. This course helps undergraduates to master the basic knowledge in the field of energy storage system detection and estimation, including the theoretical basis of energy storage detection technology and the technical basis of energy storage system sensors. Students are expected to master the theoretical basis of energy storage system estimation, including the mathematical basis of energy storage system estimation theory, and theoretical basis for system estimation. Students should master the calculation method of variance and mean square error of energy storage system detection and estimation, and master the estimation method of linear dynamic systems and the estimation method of nonlinear filtering point.

（2）Ability of analyzing the detection and estimation engineering problems for energy storage systems. This course enables undergraduates to use relevant knowledge in the field of energy storage system detection and estimation to analyze specific engineering problems in the actual applications. Students are expected to construct linear dynamic system estimation models and nonlinear filtering point estimation models for energy storage system detection and estimation in actual engineering problems, and propose effective fault detection, identification, estimation, and state estimation schemes based on the actual model characteristics.

（3）Ability to use professional knowledge combined with literature research to solve complex engineering problems. This course enables undergraduates to systematically solve the complex engineering problems in the field of energy storage system detection and estimation. Using the basic knowledge in the field of energy storage system detection and estimation, and scientific methods such as literature search and survey etc., students are expected to reasonably select the proper model system for energy storage system detection and estimation, formulating research routes, designing experiment schemes, performing experiments safely and effectively, and designing reasonable and effective fault detection, identification, estimation, and state estimation schemes for energy storage system detection and estimation problems.

7.3.2　课程基本情况（Basic Information of the Course）

课程名称	储能系统检测与估计 Detection and Estimation of Energy Storage Systems										
开课时间	一年级		二年级		三年级		四年级		总学分	2	总学时	34
	秋	春	秋	春	秋	春	秋	春				
课程定位	储能科学与工程专业本科生储能系统课程群必修课											
授课学时分配	课堂讲授32学时+综合实验2学时											
先修课程	高等数学Ⅰ,线性代数与解析几何,概率论与数理统计,自动控制理论,储能原理Ⅰ,储能原理Ⅱ											

续表

后续课程	信息物理融合能源系统
教学方式	课堂教学,大作业,实验与综述报告,平时作业
考核方式	课堂出勤占5%,课后作业占10%,课程报告占15%,课程结束笔试成绩占70%
参考教材	[1] Ludeman L C. 随机过程:滤波、估计与检测. 邱天爽,李婷,毕英伟,等,译.北京:电子工业出版社,2005.
参考资料	[1] 韩崇昭,朱洪艳,段战胜.多源信息融合. 2版. 北京:清华大学出版社,2010.

7.3.3　教学目的和基本要求（Teaching Objectives and Basic Requirements）

通过本课程的学习，使学生掌握储能系统检测与估计领域基础知识和理解能力，使学生具备储能系统检测与估计工程问题分析能力，并能够利用专业知识结合文献研究，解决复杂工程问题。

Through the study of this course, undergraduates can master the basics of detection and estimation for energy storage systems, analyze the engineering problems of energy storage system detection and estimation, and use the expertise and literature studies to solve complex engineering problems.

7.3.4　课程大纲和知识点（Syllabus and Key Points）

第一章　绪论（introduction）

章节序号 chapter number	章节名称 chapters	知识点 key points	课时 class hour
1.1	储能系统检测与估计应用背景 application background of energy storage system detection and estimation	（1）储能系统检测与估计的应用背景 （2）储能系统检测与估计的应用需求 （1）application background of energy storage system detection and estimation （2）application requirements of energy storage system detection and estimation	1
1.2	理论、方法概述及发展趋势介绍 introduction to theories, methods and development trends	（1）检测理论、方法概述 （2）估计理论、方法概述 （3）储能系统的研究现状及发展趋势 （1）introduction to theories and methods of detection （2）introduction to theories and methods of estimation （3）current status and development trends of energy storage system	1

第二章　储能系统检测技术理论基础（theoretical basis of energy storage system detection technology）

章节序号 chapter number	章节名称 chapters	知识点 key points	课时 class hour
2.1	误差分析 error analysis	（1）误差的基本概念 （2）稳态误差和动态误差 （1）basic concept of error （2）steady state errors and dynamic errors	2
2.2	测试系统的静态特性与动态特性 static and dynamic characteristics of the test system	（1）静态特性与动态特性的基本概念 （2）静态模型和动态模型 （1）basic concepts of static and dynamic characteristics （2）static and dynamic models	2

第三章　储能系统传感器技术基础（technical fundamentals of sensors in energy storage system）

章节序号 chapter number	章节名称 chapters	知识点 key points	课时 class hour
3.1	传感器的基础知识 basic knowledge of sensors	（1）传感器的定义 （2）传感器的应用 （3）传感器的分类 （4）传感器的基本特性 （5）传感器的误差 （1）definition of sensors （2）applications of sensors （3）classification of sensors （4）basic characteristics of sensors （5）errors of sensors	1
3.2	能量型传感器 energy sensor	（1）电位器式传感器 （2）电阻应变式传感器 （3）电感式传感器 （4）电容式传感器 （1）potentiometer sensor （2）resistance strain sensor （3）inductive sensor （4）capacitive sensor	1

章节序号 chapter number	章节名称 chapters	知识点 key points	课时 class hour
3.3	基于物理特性的传感器 sensors based on physical characteristics	（1）压电式传感器 （2）超声波传感器 （3）霍尔传感器 （4）光电传感器 （1）piezoelectric sensor （2）ultrasonic sensor （3）hall sensor （4）photoelectric sensor	1
3.4	环境量检测传感器 environmental quantity detection sensor	（1）气体传感器与烟雾传感器 （2）湿度传感器与水分传感器 （3）声敏传感器及超声波传感器 （1）gas sensor and smoke sensor （2）humidity sensor and moisture sensor （3）acoustic sensor and ultrasonic sensor	1

第四章 储能系统估计理论的数学基础（mathematical basis of energy storage system estimation theory）

章节序号 chapter number	章节名称 chapters	知识点 key points	课时 class hour
4.1	线性代数与线性系统理论 linear algebra and linear system theory	（1）矩阵的定义与运算 （2）矩阵的分解 （3）梯度、雅可比矩阵及黑塞矩阵 （4）连续时间线性系统及离散时间线性系统的定义与性质 （5）连续时间线性系统的离散化 （6）系统的能观能控性及其判定 （1）definition and some operations of matrix （2）matrix decomposition （3）gradient, Jacobian matrix and Hessian matrix （4）definitions and properties of continuous-time linear systems and discrete-time linear systems （5）discretization of continuous-time linear systems （6）observability and controllability of the system and its judgment	2

续表

章节序号 chapter number	章节名称 chapters	知识点 key points	课时 class hour
4.2	概率论与统计理论基础 basis of probability theory and statistics theory	（1）随机变量及其累积分布函数 （2）随机变量的概率密度函数 （3）随机变量的矩 （4）随机变量的函数变换 （5）全概率公式与贝叶斯公式 （6）高斯及联合高斯随机变量 （1）random variables and cumulative distribution function （2）probability density function of random variables （3）moments of random variables （4）function transformation of random variables （5）total probability formula and Bayes formula （6）Gaussian and joint Gaussian random variables	2

第五章 储能系统估计的理论基础（theoretical basis of energy storage system estimation）

章节序号 chapter number	章节名称 chapters	知识点 key points	课时 class hour
5.1	参数估计问题描述 description of parameter estimation problem	参数估计问题描述 description of parameter estimation problem	1
5.2	极大似然估计和最大后验估计 maximum likelihood estimation and maximum posterior estimation	（1）极大似然估计 （2）最大后验估计 （1）maximum likelihood estimation （2）maximum posterior estimation	1
5.3	最小二乘估计和最小均方误差估计 least squares estimation and minimum mean square error estimation	（1）最小二乘估计 （2）最小均方误差估计 （1）least squares estimation （2）minimum mean square error estimation	1
5.4	储能系统静态估计实例 examples of static estimation of energy storage system	（1）储能系统线性模型 （2）储能系统线性静态估计 （1）a linear model of energy storage system （2）linear static estimation of energy storage system	1

第六章 线性动态系统估计（linear dynamic system estimation）

章节序号 chapter number	章节名称 chapters	知识点 key points	课时 class hour
6.1	高斯随机变量的估计 estimation of Gaussian random variables	（1）高斯随机变量的极大似然估计 （2）高斯随机变量的最大后验估计 （1）maximum likelihood estimation of Gaussian random variables （2）maximum posterior estimation of Gaussian random variables	1
6.2	线性最小均方误差估计 linear minimum mean square error estimation	（1）正交性原理 （2）向量随机变量的线性最小均方误差估计 （1）principle of orthogonality （2）linear minimum mean square error estimation for vector random variables	2
6.3	卡尔曼滤波 Kalman filter	（1）动态估计问题 （2）卡尔曼滤波器推导 （3）卡尔曼滤波算法 （4）卡尔曼滤波的性质 （1）dynamic estimation problem （2）derivation of the Kalman filter （3）Kalman filter algorithm （4）properties of the Kalman filter	2
6.4	储能系统线性动态参数估计实例 examples of linear dynamic parameter estimation of energy storage system	（1）储能系统的线性动态模型 （2）储能系统的线性动态估计 （1）a linear dynamic model of energy storage system （2）linear dynamic estimation of energy storage system	1

第七章 非线性滤波点估计方法（nonlinear filtering point estimation method）

章节序号 chapter number	章节名称 chapters	知识点 key points	课时 class hour
7.1	非线性滤波点估计方法简介 overview of nonlinear filtering point estimation methods	（1）点估计的基本概念 （2）点估计的逼近方法 （1）basic concept of point estimation （2）approximation techniques for point estimation	0.5
7.2	扩展卡尔曼滤波 extended Kalman filter	（1）非线性系统泰勒级数展开 （2）扩展卡尔曼滤波算法 （1）Taylor series expansions of a nonlinear system （2）extended Kalman filter algorithm	0.5

续表

章节序号 chapter number	章节名称 chapters	知识点 key points	课时 class hour
7.3	无迹滤波 unscented filter	（1）无迹变换 （2）无迹滤波算法 （1）unscented transform （2）unscented filter algorithm	1
7.4	不相关转换滤波 uncorrelated conversion based filter	（1）不相关转换 （2）最优不相关转换滤波器 （1）uncorrelated conversion （2）optimized uncorrelated conversion based filter	1
7.5	储能系统非线性动态状态估计实例 examples of nonlinear dynamic state estimation of energy storage system	（1）储能系统的非线性动态模型 （2）储能系统的非线性动态状态估计 （1）nonlinear dynamic model of energy storage system （2）nonlinear dynamic state estimation of energy storage system	1

第八章　故障检测（fault detection）

章节序号 chapter number	章节名称 chapters	知识点 key points	课时 class hour
8.1	故障检测 fault detection	（1）故障的定义 （2）故障的种类 （3）故障的重构 （4）故障检测的目标 （5）故障检测的评估 （1）definitions of faults （2）types of faults （3）fault reconfiguration （4）objectives of fault detection （5）evaluation of fault detection	1
8.2	故障检测的描述 formulation of fault detection	（1）故障的描述 （2）故障的建模 （3）二元假设检验 （4）多元假设检验 （1）fault formulation （2）models of faults （3）binary hypothesis testing （4）M-ary hypothesis testing	1

续表

章节序号 chapter number	章节名称 chapters	知识点 key points	课时 class hour
8.3	故障检测算法 fault detection algorithms	（1）硬件故障检测算法 （2）基于残差的故障检测 （3）序贯故障检测算法 （4）多模型故障检测算法 （1）hardware fault detection algorithm （2）residual based fault detection algorithm （3）sequential fault detection algorithm （4）multiple-model fault detection algorithm	1
8.4	储能系统故障检测应用实例 examples of fault detection of energy storage systems	（1）基于数据的储能系统故障检测算法建模 （2）基于残差的储能系统故障检测算法 （1）data driven fault detection modeling of energy storage systems （2）based residual fault detection algorithms for energy storage systems	1

7.3.5 实验环节（Experiments）

序号 number	实验内容 experiment content	知识点 key points	课时 class hour
1	储能系统检测与估计综合实验 comprehensive experiments on detection and estimation of energy storage systems	（1）典型储能系统建模与仿真 （2）储能系统故障检测仿真实验 （3）储能系统线性动态状态估计仿真实验 （4）储能系统非线性动态状态估计仿真实验 （1）modeling and simulation of typical energy storage systems （2）simulated experiments on fault detection of energy storage systems （3）simulated experiments on linear dynamic state estimation of energy storage systems （4）simulated experiments on nonlinear dynamic state estimation of energy storage systems	2

7.4　"能源互联网"教学大纲

课程名称：能源互联网

Course Title：Internet of Energy

先修课程：运筹学，储能材料工程

Prerequisites：Operations Research，Energy Storage Material Engineering

学分：2

Credits：2

7.4.1　课程目的和基本内容（Course Objectives and Basic Contents）

本课程介绍能源互联网的发展现状、进展以及能源互联网的未来发展战略、关键技术、实践应用和机制模式。课程通过对国内外典型的能源互联网案例的介绍，如德国的 E-Energy、美国北卡罗来纳州州立大学的 FREEDM、欧洲的 FINSENY 等，使学生理解能源互联网的概念、关键技术和未来发展趋势。

The objective of this course is to introduce the state of the art of internet of energy, its future trends including development strategies, key technologies, practical applications, mechanism, and patterns. The course uses typical case studies of internet of energy in the world, e.g., Germany's E-Energy, U.S. NCSU's FREEDM, and Europe's FINSENY, etc., to help students to understand the concept of internet of energy, and associated key technologies, as well as development trends.

7.4.2　课程基本情况（Basic Information of the Course）

课程名称	能源互联网 Internet of Energy										
开课时间	一年级		二年级		三年级		四年级		总学分	2	总学时	32
	秋	春	秋	春	秋	春	秋	春				
课程定位	储能科学与工程专业本科生储能系统课程群必修课											
授课学时分配	课堂讲授32学时											
先修课程	运筹学,储能材料工程											
后续课程	电力系统分析											
教学方式	课堂教学,案例分析,讨论											
考核方式	课程论文占60%,课堂表现占30%,出勤占10%											

参考教材	[1] 能源互联网研究课题组. 能源互联网发展研究. 北京:清华大学出版社,2017. [2] 冯庆东. 能源互联网与智慧能源. 北京:机械工业出版社,2013. [3] 孙宏斌,等. 能源互联网. 北京:科学出版社,2020.
参考资料	无

7.4.3 教学目的和基本要求（Teaching Objectives and Basic Requirements）

（1）使学生了解能源互联网的基本概念，理解其中的运行、控制、保护、市场机制等问题；

（2）使学生了解储能技术的概念及其在能源互联网中的应用；

（3）使学生了解能源互联网的现状和发展趋势；

（4）对学生就业方向进行引导和帮助。

（1）Understand the concept of internet of energy, especially the operation, control, protection, and market mechanism of inernet of energy.

（2）Know the concepts of energy storage technologies and their applications in internet of energy.

（3）Know the current status and future developing trends of internet of energy.

（4）Provide guidance and help to students for job application and career development.

7.4.4 课程大纲和知识点（Syllabus and Key Points）

第一章 能源互联网中的低碳经济与灵活负荷（low-carbon economic and flexible load in the internet of energy）

章节序号 chapter number	章节名称 chapters	知识点 key points	课时 class hour
1.1	能源互联网中的低碳经济与灵活负荷 low-carbon economic and flexible load in the internet of energy	（1）低碳经济 （2）灵活负荷 （1）low-carbon economic （2）flexible load	2

第二章　全球能源互联网环境下的特高压直流输电关键技术（key technologies of ultra-high voltage direct current transmission under global internet of energy）

章节序号 chapter number	章节名称 chapters	知识点 key points	课时 class hour
2.1	全球能源互联网环境下的特高压直流输电关键技术 key technologies of ultra-high voltage direct current transmission under global internet of energy	特高压直流输电技术基本原理与关键技术 principles and key technologies of ultra-high voltage direct current transmission system	2

第三章　我国能源互联网的发展与布局（development and deployment of internet of energy in China）

章节序号 chapter number	章节名称 chapters	知识点 key points	课时 class hour
3.1	我国能源互联网的发展与布局 development and deployment of internet of energy in China	我国能源互联网的发展战略 development strategy of internet of energy in China	2

第四章　能源互联网控制与运行关键问题探讨（discussion on key problems of control and operation of internet of energy）

章节序号 chapter number	章节名称 chapters	知识点 key points	课时 class hour
4.1	能源互联网控制与运行关键问题探讨 discussion on key problems of control and operation of internet of energy	（1）能源互联网运行的基本概念 （2）能源互联网控制的基本概念 （1）concept of operation of internet of energy （2）concept of control of internet of energy	4

第五章 能源互联网中保护与控制发展新方向（new trends of protection and control of internet of energy）

章节序号 chapter number	章节名称 chapters	知识点 key points	课时 class hour
5.1	能源互联网中保护与控制发展新方向 new trends of protection and control of internet of energy	（1）电力系统保护与控制基本原理 （2）能源互联网中保护与控制面临的新技术难点 （1）concept of protection and control in power systems （2）new technical issues of protection and control in internet of energy	4

第六章 能源互联网大数据技术与信息物理融合技术（techniques of big data and cyber-physical system in internet of energy）

章节序号 chapter number	章节名称 chapters	知识点 key points	课时 class hour
6.1	能源互联网大数据技术与信息物理融合技术 techniques of big data and cyber-physical system in internet of energy	（1）大数据技术基本概念及其在能源互联网中应用 （2）信息物理融合系统概念及其在能源互联网中应用 （1）concept of big data techniques and its application in internet of energy （2）concept of cyber-physical system and its application in internet of energy	6

第七章 能源互联网中储能技术及其应用（energy storage technologies and their applications in internet of energy）

章节序号 chapter number	章节名称 chapters	知识点 key points	课时 class hour
7.1	能源互联网中储能技术及其应用 energy storage technologies and their applications in internet of energy	（1）储能技术的概念和基本原理 （2）储能技术在能源互联网中的应用 （1）concept and principles of energy storage （2）application of energy storage technologies in internet of energy	8

第八章　能源互联网的市场机制和商业模式（market mechanism and business model in internet of energy）

章节序号 chapter number	章节名称 chapters	知识点 key points	课时 class hour
8.1	能源互联网的市场机制和商业模式 market mechanism and business model in internet of energy	（1）能源市场的基本概念，包括市场主体、交易机制、定价等基本原理 （2）能源互联网中的投资模式、运营模式、政策影响等的基本概念和实例 （1）concept of energy market, including the principles on market entities, transaction mechanism, and pricing, etc. （2）concept and examples of investment model, operation model, and impact of policies in internet of energy	4

7.5　"嵌入式智能系统"教学大纲

课程名称：嵌入式智能系统

Course Title：Embedded Intelligent System

先修课程：电路，计算机科学基础与高级程序设计

Prerequisites：Electric Circuit, Computer Science Fundamentals and Advanced Programming Design

学分：2

Credits：2

7.5.1　课程目的和基本内容（Course Objectives and Basic Contents）

本课程使学生掌握嵌入式系统的组成、工作原理及设计方法，从而具备计算机硬件系统的分析和设计能力，以适应工程领域高级技术人才的需求，并支撑毕业要求中的相应指标点。

课程讲述 ARM 系列微处理器体系结构、指令系统。在此基础上，学习使用以 ARM 系列微处理器为内核的 SOC 芯片设计典型的嵌入式系统，并进一步学习在实时 Linux 系统下的编程方法。

This course enables students to master the organization, working principle, and design method of embedded systems, to analyze and design computer hardware systems, which should meet the needs of senior technical personnel in the engineering field, and support the corresponding index points in graduation requirements.

The course focuses on the architecture and instruction system of ARM series microprocessor. On this basis, it introduces how to use SOC chip with ARM core to design typical embedded systems, and further introduces how to program under real-time Linux system.

7.5.2　课程基本情况（Basic Information of the Course）

课程名称	嵌入式智能系统 Embedded Intelligent System									
开课时间	一年级		二年级		三年级		四年级		总学分	2	总学时	44
	秋	春	秋	春	秋	春	秋	春				
课程定位	储能科学与工程专业本科生储能系统课程群必修课									
授课学时分配	课堂讲授36学时+实验8学时									
先修课程	电路,计算机科学基础与高级程序设计									
后续课程	储能装置设计与开发实验									
教学方式	课堂教学,平时作业,实验,讨论									
考核方式	课程结束笔试成绩占60%,实验报告占20%,平时成绩占20%									
参考教材	[1] 常华,黄岚,张海燕.嵌入式系统原理与应用.北京:清华大学出版社,2013.									
参考资料	[1] 张晨曦,韩超,沈立,等.嵌入式系统教程.北京:清华大学出版社,2013.									

7.5.3　教学目的和基本要求（Teaching Objectives and Basic Requirements）

（1）掌握嵌入式系统的基本组成和工作原理，能够熟练地对其原理或过程进行描述，建立"整机"概念。

（2）熟悉嵌入式系统的发展历史和在储能领域的应用现状。

（3）能够准确地分析工程领域中嵌入式系统的设计需求，明确设计目标与任务，提出综合满足成本、功耗、体积、性能等要求的嵌入式系统设计方案，包括硬件设计及软件编程等；并通过仿真、实验等手段论证方案的正确性和可行性。在此基础上对设计方案进行优化和改进。

（1）Master the basic organization and working principle of embedded system, be able to describe its principle or process skillfully, and establish the concept of "the whole machine".

（2）Be familiar with the development of embedded systems and their applications in energy storage field.

（3）Accurately analyze the design requirements of embedded systems in the engineering field, clarify the design objectives and tasks, put forward the embedded system design scheme, including hardware design and software programming, to comprehensively meet the requirements of cost, power consumption, volume, and performance. Demonstrate the correctness and feasibility of the scheme by means of simulation and experiment. On this basis, further optimize and improve the designed scheme.

7.5.4 课程大纲和知识点（Syllabus and Key Points）

第一章 嵌入式系统简介（introduction of embedded systems）

章节序号 chapter number	章节名称 chapters	知识点 key points	课时 class hour
1.1	嵌入式系统的定义、特点及架构 definition, characteristics, and architecture of embedded systems	（1）嵌入式系统的定义 （2）嵌入式系统的特点 （3）嵌入式系统的架构 （4）嵌入式微处理器 （1）definition of embedded systems （2）characteristics of embedded systems （3）architecture of embedded systems （4）embedded microprocessor	0.5
1.2	嵌入式系统的应用领域及发展趋势 application and development of embedded systems	（1）应用领域 （2）发展趋势 （1）applications （2）developments	0.2
1.3	嵌入式系统的设计流程 design flow of embedded system	嵌入式系统的设计流程 design flow of embedded system	0.3

第二章 计算机组成原理基础（fundamentals of computer organization）

章节序号 chapter number	章节名称 chapters	知识点 key points	课时 class hour
2.1	计算机组成 computer organization	（1）冯·诺依曼计算机 （2）现代计算机的特点 （1）von Neumann machine （2）characteristics of modern computer	1
2.2	计算机工作原理 working principle of computer	工作原理 working principle	1

第三章 ARM 嵌入式微处理器体系结构（architecture of ARM embedded microprocessor）

章节序号 chapter number	章节名称 chapters	知识点 key points	课时 class hour
3.1	ARM体系结构概览 overview of ARM architecture	（1）精简指令系统计算机 （2）ARM/THUMB指令集 （3）ARM的指令流水线 （4）哈佛结构 （1）RISC （2）ARM/THUMB instruction set （3）instruction pipeline of ARM （4）Harvard structure	2
3.2	ARM9编程模型 ARM9 programming model	（1）ARM的两种操作状态 （2）7种工作模式 （3）ARM的通用寄存器 （1）two operating states （2）7 working modes （3）general registers	2
3.3	ARM9异常处理 ARM9 exception handling	（1）异常（中断）类型 （2）异常响应流程 （3）异常返回 （1）exception (interrupt) types （2）exception response process （3）exception return	3

第四章 ARM 嵌入式处理器的指令系统（instruction system of ARM）

章节序号 chapter number	章节名称 chapters	知识点 key points	课时 class hour
4.1	ARM指令集特点 characteristics of ARM instruction set	（1）特点 （2）语法规则 （1）characteristics （2）rules of instruction grammar	1
4.2	ARM的寻址方式 addressing modes of ARM	（1）基址变址寻址 （2）多寄存器寻址 （3）堆栈寻址 （1）based indexed addressing （2）multi register addressing （3）stack addressing	2

续表

章节序号 chapter number	章节名称 chapters	知识点 key points	课时 class hour
4.3	ARM指令集详解 details of ARM instruction set	（1）控制流指令 （2）load/store指令 （3）数据处理指令 （4）软中断指令 （1）control flow instruction （2）load/store instruction （3）data processing instruction （4）SWI instruction	2
4.4	ARM伪指令 pseudo instruction	（1）变量定义伪指令 （2）ARM伪指令 （1）variable definition directive （2）ARM pseudo instruction	2

第五章 ARM 的编程（ARM programming）

章节序号 chapter number	章节名称 chapters	知识点 key points	课时 class hour
5.1	开发工具 development tool	（1）开发工具 （2）映像文件 （1）development tool （2）image file	2
5.2	ARM-Thumb子程序调用标准（ATPCS） ARM-Thumb Procedure Call Standard	（1）规则 （2）举例 （1）rules （2）examples	4

第六章 ARM 最小系统设计（design of ARM minimum system）

章节序号 chapter number	章节名称 chapters	知识点 key points	课时 class hour
6.1	S3C2440及最小系统 S3C2440 and minimum system	（1）特点 （2）语法规则 （1）characteristics （2）rules of instruction grammar	1

续表

章节序号 chapter number	章节名称 chapters	知识点 key points	课时 class hour
6.2	最小系统的设计 design of minimum system	（1）SoC的概念 （2）S3C2440介绍 （3）最小系统的设计 （1）concept of SoC （2）introduction to S3C2440 （3）design of minimum system	3

第七章 ARM 系统的设计（design of ARM system）

章节序号 chapter number	章节名称 chapters	知识点 key points	课时 class hour
7.1	系统时钟及电源管理 clock and power management	（1）系统时钟管理 （2）电源管理 （1）system clock management （2）power management	1
7.2	GPIO设计 design of GPIO	（1）GPIO介绍 （2）GPIO使用举例 （1）introduction to GPIO （2）examples of how to use GPIO	2
7.3	中断系统设计 design of interrupt system	（1）程序框架 （2）S3C2440中断源的组织 （3）外部中断服务程序框架 （1）program framework （2）organization of interrupt source in S3C2440 （3）external interrupt routine framework	3
7.4	通用异步收发器UART universal Asynchronous Receiver/Transmitter	（1）原理 （2）UART使用举例 （1）principle （2）examples of how to use UART	2

第八章 Boot Loader 技术及 Linux 编程（Boot Loader technology and Linux programming）

章节序号 chapter number	章节名称 chapters	知识点 key points	课时 class hour
8.1	Boot Loader技术 Boot Loader technology	Boot Loader技术 Boot Loader technology	0.5

续表

章节序号 chapter number	章节名称 chapters	知识点 key points	课时 class hour
8.2	Linux编程 Linux programming	Linux编程 Linux programming	0.5

7.5.5　实验环节（Experiments）

序号 number	实验内容 experiment content	知识点 key points	课时 class hour
1	汽车防盗系统 car anti-theft system	（1）开发板原理 （2）SDRAM系统 （3）Flash系统 （4）网络及GPRS接口 （5）日历时钟电路 （6）振动检测 （1）principle of development board （2）SDRAM system （3）Flash system （4）network and GPRS interface （5）calendar clock circuit （6）vibration detection	8

7.6　"智能电网储能应用技术"教学大纲

课程名称：智能电网储能应用技术

Course Title：Applications of Energy Storage Technology in Smart Grid

先修课程：电路，电力系统分析

Prerequisites：Electric Circuit, Power Systems Analysis

学分：2

Credits：2

7.6.1　课程目的和基本内容（Course Objectives and Basic Contents）

通过本课程的学习，使学生掌握智能电网储能应用的基本概念、基本理论和基本规律，并运用系统工程理论、观点和方法，分析、研究、计算或估计一些典型储能应用问题，从而培养学生科学的思想方法和工程问题的解决能力。

本课程主要包括储能系统运行控制技术、储能技术发电侧应用、储能技术用户侧应用、储能对电力系统规划运行的影响、储能参与电力市场的商业模式等内容。

Through this course, students can master the basic concepts, theories, and laws of energy storage applications in smart grids. Students can learn how to analyze, study, calculate, or estimate some typical energy storage application problems through using system engineering theories, viewpoints and methods. This course is very helpful to cultivate students' scientific thinking and ability for solving engineering problems.

The course focuses on operational controls of energy storage system, applications of energy storage in the generation side, applications of energy storage in the customer side, impacts of energy storage on power system planning and operation, and business models of energy storage in the electricity market.

7.6.2 课程基本情况（Basic Information of the Course）

课程名称	智能电网储能应用技术 Applications of Energy Storage Technology in Smart Grid								
开课时间	一年级		二年级		三年级		四年级		总学分	2	总学时	36
	秋	春	秋	春	秋	春	秋	春				

课程定位	储能科学与工程专业本科生储能系统课程群选修课
授课学时分配	课堂讲授32学时+实验4学时
先修课程	电路,电力系统分析
后续课程	储能装置设计与开发实验
教学方式	课堂教学,实验指导,平时作业,实验报告
考核方式	课程结束笔试成绩占70%,实验报告占20%,平时成绩占10%
参考教材	[1] 吴福保,杨波,叶季蕾.电力系统储能应用技术.北京:中国水利水电出版社,2014. [2] 孙威,李建林,王明旺.能源互联网:储能系统商业运行模式及典型案例分析.北京:中国电力出版社,2017.
参考资料	[1] 王松岑.大规模储能技术及其在电力系统中的应用.北京:中国电力出版社,2016. [2] 唐西胜,齐智平,孔力.电力储能技术及应用.北京:机械工业出版社,2020.

7.6.3 教学目的和基本要求（Teaching Objectives and Basic Requirements）

教学目的：

（1）使学生了解各类电储能类型、技术适应性等系统知识；

（2）培养学生掌握储能电池管理系统的原理及分析方法；

（3）培养学生认识发电侧的储能应用技术并掌握分析方法；

（4）培养学生认识用户侧的储能应用技术并掌握分析方法；

（5）培养学生了解储能对电力系统规划运行影响的物理概念，并掌握储能的电力系统规划运行分析方法；

（6）培养学生了解储能参与市场的技术并掌握分析方法。

基本要求：

（1）掌握储能的类型和技术适用性，了解储能的典型应用场景和示范工程；

（2）掌握储能电池管理系统的基本原理，掌握储能系统的运行控制技术；

（3）掌握储能技术在新能源并网发电中的应用以及在常规发电中的应用；

（4）掌握储能技术在微电网中的相关应用，掌握用户侧分布式储能技术；

（5）掌握计及储能的电力系统规划方法，掌握计及储能的电力系统优化运行技术，了解储能对电力系统暂态运行的影响；

（6）掌握储能参与电力市场的相关技术，了解储能技术的商业模式。

Educational purposes:

（1）Understand the systematic knowledge of various types of electric energy storage and technical adaptability.

（2）Understand the principles and analysis methods of energy storage battery management system and master the analysis method.

（3）Understand the energy storage application technology on the power generation side and master the corresponding analysis method.

（4）Understand the energy storage application technology on the customer side and master the corresponding analysis method.

（5）Understand the physical concept of energy storage's influence on power system planning and operation, and to master the analysis method of power system planning and operation considering energy storage.

（6）Understand the technology of the energy storage system participating in the electricity market and master the corresponding analysis method.

Basic Requirements:

（1）Master energy storage types and technical applicability, and understand typical application scenarios and demonstration projects of energy storage system.

（2）Master the basic principles of energy storage battery management system and the operation control technology of energy storage system.

（3）Master the application of energy storage technology in grid-connected renewable energy generation and the application of energy storage technology in the conventional power generation.

（4）Master the application of energy storage technology in microgrid and the distributed energy storage technology in the customer side.

（5）Master the power system planning method including energy storage system, master the optimal operation technology of power system including energy storage system, and understand the influence of energy storage on power system transients.

（6）Master the related technologies of energy storage participating in electricity market, and understand the business model of energy storage technology.

7.6.4 课程大纲和知识点（Syllabus and Key Points）

章节序号 chapter number	章节名称 chapters	知识点 key points	课时 class hour
1	绪论 introduction	（1）储能在智能电网中的作用 （2）储能的类型和技术适用性 （3）储能的典型应用场景和示范工程 （1）applications of energy storage in smart grid （2）types and technical applicability of energy storage （3）application cases and demonstration projects	4
2	储能系统运行控制技术 operational control technology of energy storage system	（1）储能电池管理系统原理 （2）储能系统运行控制技术 （1）basic principle of energy storage battery management system （2）operational control techniques of energy storage system	6
3	储能技术发电侧应用 applications of energy storage in generation side	（1）在新能源并网发电中的应用 （2）在常规发电中的应用 （1）applications of energy storage system in grid-connected renewable energy generation （2）applications of energy storage system in the conventional electricity generation	4
4	储能技术用户侧应用 applications of energy storage in customer side	（1）微电网的储能应用 （2）分布式储能技术 （1）applications of energy storage system in microgrids （2）distributed energy storage technology	6
5	储能对电力系统规划运行的影响 impacts of energy storage on power system planning and operation	（1）电力系统的储能投资规划 （2）含储能的电力系统的优化运行 （3）储能对电力系统的暂态影响 （1）energy storage investment planning in power system （2）optimal operations of power systems with energy storage （3）impacts of energy storage on power system transients	6

续表

章节序号 chapter number	章节名称 chapters	知识点 key points	课时 class hour
6	储能参与电力市场的商业模式 business model of energy storage participating in electricity market	（1）储能参与电力市场 （2）储能的商业模式 （1）energy storage participating in electricity market （2）business model of energy storage	6

7.6.5　实验环节（Experiments）

序号 number	实验内容 experiment content	知识点 key points	课时 class hour
1	新能源与储能联合优化运行 optimal operation of renewable energy resources combined with energy storage systems	（1）新能源独立运行特性 （2）储能运行特性 （3）风电场-储能优化运行 （4）光伏电站-储能优化运行 （1）operating characteristics of renewable energy resources （2）operating characteristics of energy storage （3）optimal operation of wind power and energy storage （4）optimal operation of photovoltraic power and energy storage	2
2	含储能的微网能量管理 energy management system for microgrids with energy storages	（1）微网各元件运行特性 （2）并网与离网运行模式 （3）微网能量管理策略 （4）微网整体优化运行 （1）operating characteristics of microgrid components （2）microgrid operations in offline and online modes （3）strategic energy management for microgrids （4）optimal operation of microgrids	2

7.7　"电储能系统与并网技术"教学大纲

课程名称：电储能系统与并网技术

Course Title：Electric Energy Storage System and Grid Connection Technology

先修课程：电路，现代电子技术

Prerequisites：Electric Circuit, Modern Electronic Technology

学分：2

Credits：2

7.7.1 课程目的和基本内容（Course Objectives and Basic Contents）

本课程是储能科学与工程专业本科生的一门选修课。本课程围绕电储能系统的基本概念、电功率转换系统及电储能并网控制技术的现状和研究发展趋势展开。

（1）使学生对电储能系统有全面的了解，对其特性、作用、相互间的关系能用全局的观点去认识，学会电储能系统的相关理论及综合分析工程问题的基本方法；

（2）使学生进一步巩固、理解先修课程有关的理论和方法，并掌握这些理论和方法在"电储能系统与并网技术"课程中的应用；

（3）训练学生工程计算能力，培养学生分析和解决电储能系统工程问题的能力；

（4）掌握运用功率电能变换单元的理论和方法；

（5）了解电储能系统的发展趋势及其发展新技术，提高学生自主学习意识。

课程以电储能系统的概念、电储能变换系统、功率半导体器件、电储能变换技术、电储能并网技术及应用为核心。

This course is an elective course for undergraduates majoring in Energy Storage Science and Engineering. This course focuses on the basic concepts of electric energy storage systems, the current status, and research development trends of electric power conversion systems and grid-connected control technologies.

（1）Enable students to have a comprehensive understanding of the electrical energy storage system, to understand its characteristics, functions, and mutual relationships from a global perspective, and learn to comprehensively analyze the basic methods of electrical energy storage system related theories and engineering issues.

（2）Enable students to further consolidate and understand the theories and methods related to the prerequisite courses, and master the application of these theories and methods in the course "Electric Energy Storage System and Grid Connection Technology".

（3）Develop skills of students in engineering calculation, principle analysis and solution design of engineering problems of electric energy storage systems.

（4）Enable students to familiar with fundamental of power conversion system and related methods.

（5）Understand the development trend of electric energy storage systems, and the development of new technologies to improve learning awareness of students.

The course focuses on power electronics fundamental concept, structure, power device, conversion technology, and application of electric energy storage systems.

7.7.2　课程基本情况（Basic Information of the Course）

课程名称	电储能系统与并网技术 Electric Energy Storage System and Grid Connection Technology									
开课时间	一年级		二年级		三年级		四年级		总学分	2	总学时	40
	秋	春	秋	春	秋	春	秋	春				
课程定位	储能科学与工程专业本科生储能系统课程群选修课											
授课学时分配	课堂讲授32学时+实验8学时											
先修课程	电路，现代电子技术											
后续课程	无											
教学方式	课堂教学，大作业，实验与综述报告，讨论，平时作业											
考核方式	课程结束笔试成绩占70%，实验报告占20%，平时成绩占10%											
参考教材	[1] 蔡旭，李睿，李征. 储能功率变换与并网技术.北京：科学出版社，2019.											
参考资料	无											

7.7.3　教学目的和基本要求（Teaching Objectives and Basic Requirements）

通过本课程的学习，使学生了解电储能系统的概念，熟悉电储能系统中的器件与拓扑，掌握控制策略、调制策略以及并网控制策略的基本原理，了解目前储能系统的应用情况。

基本要求：

（1）掌握电储能系统的概念和内容；

（2）熟悉电储能系统中的器件、拓扑、控制策略、调制策略，以及并网控制策略的基本原理、发展历史、应用现状。

Master the concept of electric energy storage systems. Master the operation principle of device, topology, control strategy and modulation strategy of electric energy storage systems. Master control strategy for island mode and grid-connected mode. Understand the development trend, applications of energy storage system.

Basic Requirements：

（1）Master the concept of electric energy storage systems.

（2）Be familiar with the device, topology, control strategy of power converter for island and grid-connected modes.

7.7.4　课程大纲和知识点（Syllabus and Key Points）

第一章　绪论（introduction）

章节序号 chapter number	章节名称 chapters	知识点 key points	课时 class hour
1.1	电储能系统 electric energy storage system	电储能系统的概念 concept of electric energy storage system	0.5
1.2	电能变换系统 electric power conversion system	电能变换技术的概念 concept of electric power conversion	0.5
1.3	电能变换技术 electric power conversion technology	电能变换技术的应用 applications of electric power conversion technology	0.5
1.4	课程的任务与要求 tasks and requirements of the course	本课程的任务与要求 tasks and requirements of the course	0.5

第二章　电储能变换的器件介绍（introduction of power devices of electric energy storage system）

章节序号 chapter number	章节名称 chapters	知识点 key points	课时 class hour
2.1	电力电子器件 power devices	电力电子器件概述 introduction power devices	1
2.2	电力二极管 power diode	电力二极管的结构、特点及开关方式 structure, characteristic, and control of power diode	1
2.3	典型全控型器件 typical power switches	典型全控型器件的结构、特点及开关方式 structure, characteristic, and control of typical power switches	1
2.4	其他新型电力电子器件 other power switches and power modules	其他新型电力电子器件的结构、特点及开关方式 structure, characteristic, and control of other power switches and modules	1

第三章　电储能变换电路拓扑（topologies of electric energy storage circuits）

章节序号 chapter number	章节名称 chapters	知识点 key points	课时 class hour
3.1	整流电路 rectifier circuits	电路的运行机理和输出特性 principle and output characteristic of circuit	2
3.2	逆变电路 inverter circuits	电路的运行机理和输出特性 principle and output characteristic of circuit	2
3.3	直流-直流变流电路 DC-DC circuits	电路的运行机理和输出特性 principle and output characteristic of circuit	2

第四章　电能变换控制技术（control strategies of electric energy storage system）

章节序号 chapter number	章节名称 chapters	知识点 key points	课时 class hour
4.1	PWM控制的基本原理 fundamental of PWM control	PWM的概念和分类 concept and classification of PWM control	1.5
4.2	PWM逆变电路及其控制方法 PWM control in inverter circuit	PWM逆变电路及其控制方法运行机理和输出特性 principle and output characteristic of PWM control in inverter system	1.5
4.3	PWM跟踪控制技术 PWM tracking control methods	PWM跟踪控制技术的运行机理和输出特性 principle and output characteristic of PWM tracking control methods	1.5
4.4	PWM整流电路及其控制方法 PWM control in rectifier circuits	PWM整流电路及其控制方法的运行机理和输出特性 principle and output characteristic of PWM control in rectifier system	1.5

第五章　电储能系统中的电能变换应用及设计（design and application of electric energy storage system）

章节序号 chapter number	章节名称 chapters	知识点 key points	课时 class hour
5.1	电储能系统概述 review of electric energy storage system	电储能系统概述 review of electric energy storage system	1.5

续表

章节序号 chapter number	章节名称 chapters	知识点 key points	课时 class hour
5.2	电储能功率转换系统 conversion of electric energy storage system	电储能功率转换系统的运行机理及输出特性 principle and output characteristic of conversion of electric energy storage system	1.5
5.3	典型电储能系统的电路设计 typical design of electric energy storage circuit	典型电储能系统的电路设计方法及步骤 method and design steps of typical design of electric energy storage circuit	1.5
5.4	典型电储能系统的控制设计 typical design of electric energy storage control system	典型电储能系统的控制设计方法及步骤 method and design steps of typical design of electric energy storage control system	1.5

第六章 电储能系统的并离网技术（technology of grid-connected to off-grid electric energy storage system）

章节序号 chapter number	章节名称 chapters	知识点 key points	课时 class hour
6.1	电储能系统的并网控制策略 grid-connected control strategy of electric energy storage control system	电储能系统的并网控制策略分析及参数设计 analysis and parameter design of grid-connected electric energy storage control system	2
6.2	电储能系统的离网控制策略 island control strategy of electric energy storage control system	电储能系统的离网控制策略分析及参数设计 analysis and parameter design of island electric energy storage control system	2

第七章 电储能的工程示范应用（applications of electric energy storage system）

章节序号 chapter number	章节名称 chapters	知识点 key points	课时 class hour
7.1	大容量电储能示范工程 demonstration projects of electric energy storage system	大容量电储能示范工程 demonstration projects of electric energy storage system	2

续表

章节序号 chapter number	章节名称 chapters	知识点 key points	课时 class hour
7.2	光储一体化示范工程 demonstration projects of PV and electric energy storage system	光储一体化示范工程 demonstration projects of PV and electric energy storage system	1
7.3	风储一体化示范工程 demonstration projects of wind and electric energy storage system	风储一体化示范工程 demonstration projects of wind and electric energy storage system	1

7.7.5　实验环节（Experiments）

序号 number	实验内容 experiment content	知识点 key points	课时 class hour
1	电储能系统控制性能研究 control of electric energy storage system	（1）熟悉实验平台的电路结构和主要元器件，连接实验线路 （2）接通控制和驱动电源，将载波频率设置为 10 kHz，控制方式为开环、充电模式，测量电储能系统的电压、电流波形 （3）接通控制和驱动电源，将载波频率设置为 20 kHz，控制方式为闭环、放电模式，测量电储能系统的电压、电流波形 （1）study the circuit structure and main components of the experimental platform, and connect the experimental circuit （2）switch on the control circuit and drive circuit, set the carrier frequency to 10 kHz, measure the voltage and current waveforms of the electric energy storage system in open-loop, charging mode （3）switch on the control and drive power supply, set the carrier frequency to 20 kHz, measure the voltage and current waveforms of the electric energy storage system in closed-loop, discharging mode	4

续表

序号 number	实验内容 experiment content	知识点 key points	课时 class hour
2	光储一体系统性能研究 control of PV and electric energy storage system	（1）熟悉实验平台的电路结构和主要元器件,连接实验线路 （2）接通光伏控制和驱动电源,将载波频率设置为10 kHz,控制方式为开环调节光伏电压模式,测量光伏板的电压、电流波形 （3）接通储能控制和驱动电源,将载波频率设置为10 kHz,控制方式为开环、充电模式,测量电储能系统的电压、电流波形 （4）将光储系统互联,实现光伏发电向电储能充电模式,将载波频率设置为10 kHz,控制方式为开环、充电模式,测量电储能系统的电压、电流波形 （1）study the circuit structure and main components of the experimental platform, and connect the experimental circuit （2）turn on the control circuit and drive circuit, set the carrier frequency to 10 kHz, the control mode is the open-loop voltage control mode, and measure the voltage and current waveforms of the photovoltaic panel （3）switch on the energy storage control circuit and drive circuit supply, set the carrier frequency to 10 kHz, the control mode is open loop, charging mode, and measure the voltage and current waveforms of the energy storage system （4）interconnect the optical storage system to realize the charging mode from photovoltaic power , set the carrier frequency to 10 kHz, the control mode is open loop, charging mode, and measure the voltage and current waveforms of the electric energy storage system	4

7.8 "物联网应用概论"教学大纲

课程名称: 物联网应用概论

Course Title: The Application of Internet of Things

先修课程: 高等数学Ⅰ,计算机科学基础与高级程序设计

Prerequisites: Advanced Mathematics Ⅰ , Computer Science Fundamentals and Advanced Programming Design

学分：2.0

Credits：2.0

7.8.1　课程目的和基本内容（Course Objectives and Basic Contents）

本课程是储能科学与工程专业的一门重要的专业选修课，其目的是使学生掌握物联网的基本概念、了解物联网的发展现状、掌握物联网的关键技术，并通过典型案例的学习，使学生对物联网及其应用有较清晰的认识，为将来从事相关工作打下一定的基础。

课程以 6 个章节为基本内容，包括物联网的概念与关键技术、物联网感知技术、物联网标识技术、物联网传输技术、物联网数据处理与安全技术、物联网的综合应用。

This course is an important elective course for students majoring in energy storage science and engineering. The purpose is to enable students to master the basic concepts of the internet of things(IoT), understand the development status of IoT, and master the key technologies of IoT. Through the study of its typical application fields and cases, students will have a clearer understanding of IoT and its applications, which lays a certain foundation for future research and application of IoT.

The course focuses on 6 chapters, including concepts and key technologies of IoT, IoT sensing technology, IoT identification technology, IoT transmission technology, IoT data processing and security technology, and comprehensive applications of IoT.

7.8.2　课程基本情况（Basic Information of the Course）

课程名称	物联网应用概论 The Application of Internet of Things										
开课时间	一年级		二年级		三年级		四年级		总学分	2	总学时	32
	秋	春	秋	春	秋	春	秋	春				
课程定位	储能科学与工程专业本科生储能系统课程群选修课											
授课学时分配	课堂讲授32学时											
先修课程	高等数学Ⅰ,计算机科学基础高级程序设计											
后续课程	无											
教学方式	课堂教学,大作业,讨论,平时作业											
考核方式	课程结束笔试成绩占60%,平时成绩占40%											
参考教材	[1] 桂小林,安健.物联网技术原理.北京:高等教育出版社,2016. [2] 桂小林.物联网技术概论.2版.北京:清华大学出版社,2018.											
参考资料	[1] 黄玉兰.物联网射频识别核心技术详解.北京:人民邮电出版社,2010.											

7.8.3 教学目的和基本要求（Teaching Objectives and Basic Requirements）

（1）掌握物联网的相关概念、关键技术和典型应用，包括物联网体系架构、典型传感器检测原理、无线传感网络、条形码与 RFID 识别技术、短距离无线通信技术、物联网典型应用等，综合应用于物联网工程设计与实施等。

（2）掌握物联网工程设计的分析方法、实验技能，能够通过文献查阅、分析或实验、实践，对物联网工程问题的影响因素和关键环节（要素）等进行分析鉴别。能够对工程设计与实施问题中所涉及的物理现象、网络原理以及应用软件进行初步的理论分析或实验测试、验证。

（3）引导学生应用物联网相关技术及软件进行物联网应用系统的需求分析、方案设计，逐步具有应用软件工具解决工程实际问题的能力。

（4）基于课程学习，使学生理解物联网工程技术与社会发展的相互关系，树立全面客观的工程社会意识观，在工程实践中理解并遵守工程职业道德和规范。

（5）能够通过自主学习、查阅资料等方式，具备了解物联网工程领域技术和观念发展、变化的能力，为以后从事物联网行业相关研究和应用工作打下良好的基础。

（1）Master the related concepts, key technologies, and typical applications of IoT, including the architecture of the IoT, typical sensor detection principles, wireless sensor networks, barcode and RFID recognition technology, short-range wireless communication technology, and typical applications of IoT, etc. And apply the relevant technologies to the engineering design and implementation of IoT.

（2）Master the analysis methods and experimental skills of IoT engineering design and be able to analyze and identify the influencing factors and key links (elements) of IoT engineering issues through literature review, analysis, experimentation, and practice.

（3）Guide students to apply the related technologies and software to conduct demand analysis and program design of the application system of IoT, and have the ability to apply software tools to solve practical engineering problems.

（4）Based on the course study, students can understand the relationship between IoT engineering technology and social development, establish a comprehensive and objective view of engineering social consciousness, and understand and abide by engineering professional ethics and norms in engineering practice.

（5）Understand the field technologies, concept development and the ability of IoT through self-learning and literature review. And lay a good foundation for future research and application work related to the IoT industry.

7.8.4 课程大纲和知识点（Syllabus and Key Points）

第一章 物联网的概念与关键技术（concepts and key technologies of IoT）

章节序号 chapter number	章节名称 chapters	知识点 key points	课时 class hour
1.1	物联网技术介绍 introduction of IoT technology	本课程研究的对象、内容 object and content of this course	1
1.2	物联网基本概念与发展历程 basic concepts and development history of IoT	物联网技术的发展现状 development status of IoT technology	1
1.3	物联网体系结构及关键技术 IoT architecture and key technologies	物联网技术的学习方法 learning method of IoT technology	2

第二章 物联网感知技术（IoT sensing technology）

章节序号 chapter number	章节名称 chapters	知识点 key points	课时 class hour
2.1	物联网感知与识别技术概述 overview IoT sensing and recognition technology	物联网传感与检测技术 IoT sensing and detection technology	1
2.2	典型传感器原理 typical sensor principle	电阻应变式、电感式、电容式传感器检测原理 detection principle of resistance strain type, inductive type, and capacitive sensor	2
2.3	无线传感网 wireless sensor network	无线传感网及组网技术 wireless sensor network and networking technology	2
2.4	典型案例分析 typical case analysis	传感器及无线传感网典型应用 typical applications of sensors and wireless networks	1

第三章 物联网标识技术（IoT identification technology）

章节序号 chapter number	章节名称 chapters	知识点 key points	课时 class hour
3.1	条形码技术 barcode technology	一维码、二维码识别原理 recognition principle of one-dimensional code and two-dimensional code	3
3.2	RFID技术 RFID technology	RFID技术原理 principle of RFID technology	3
3.3	定位技术 positioning technology	物联网定位技术 IoT positioning technology	1
3.4	室内定位技术 indoor positioning technology	物联网室内定位技术 IoT indoor positioning technology	1
3.5	典型案例分析 typical case analysis	典型应用 typical applications	2

第四章 物联网传输技术（IoT transmission technology）

章节序号 chapter number	章节名称 chapters	知识点 key points	课时 class hour
4.1	物联网短距离无线通信技术 IoT short-range wireless communication technology	典型短距离无线通信技术 typical short-range wireless communication technologies	1
4.2	无线宽带网 wireless broadband network	WiFi协议及其组网技术 WiFi protocol and networking technology	1
4.3	无线低速网 wireless low speed network	ZigBee组网技术 ZigBee networking technology.	1
4.4	移动通信网 mobile communication network	移动蜂窝网络 Mobile cellular network	1
4.5	典型案例分析 typical case analysis	典型应用 typical applications	

第五章　物联网数据处理与安全技术（data processing and security technology of IoT）

章节序号 chapter number	章节名称 chapters	知识点 key points	课时 class hour
5.1	物联网数据管理与服务技术 management and service technology of IoT	物联网数据处理与挖掘 IoT data processing and mining	1
5.2	大数据与海量存储 big data and mass storage	物联网大数据处理技术 IoT big data processing technology	1
5.3	云计算技术 cloud computing technology	云计算架构及应用模式 cloud computing architecture and application models	1
5.4	物联网信息安全与隐私保护技术 information security and privacy protection technology of IoT	K-means、差分等隐私保护术 privacy protection technologies such as K-means and differential privacy, etc.	1

第六章　物联网综合应用（comprehensive applications of IoT）

章节序号 chapter number	章节名称 chapters	知识点 key points	课时 class hour
6.1	物联网应用系统设计 design of IoT application system	（1）物联网工程需求分析技术 （2）物联网方案设计与实施流程 （1）requirements analysis technology of IoT engineering （2）design and implementation process of IoT	2
6.2	智慧城市 smart city	智慧城市 mart city	2

7.9　"信息物理融合能源系统"教学大纲

课程名称：信息物理融合能源系统

Course Title：Cyber-Physical Energy Systems

先修课程：运筹学，自动控制理论，储能系统检测与估计

Prerequisites：Operations Research, Principles of Automatic Control, Detection and Estimation of Energy Storage Systems

学分：2

Credits：2

7.9.1　课程目的和基本内容（Course Objectives and Basic Contents）

　　本课程围绕新型感知、传输、计算等信息技术应用于能源系统的优化和安全的主题，介绍通过可再生能源、可再生能源与传统能源、能源生产与需求等之间的配合协调，实现能源电力系统的安全节能优化的系统化方案。

　　课程主要内容包括信息物理融合系统、信息物理融合能源系统的典型结构、信息物理融合能源系统的信息感知、含新能源的电力系统发电优化调度、企业能源系统与楼宇能源系统的控制与优化、信息物理融合系统综合安全。

　　This course focuses on the emerging topics on sensing, transmission, computation, and other information technologies applied to the optimization and security of energy systems. It introduces a systematic optimal solution on energy saving and security through the coordination among renewable energy, traditional generators, energy demand, etc.

　　The main contents of the course include cyber-physical systems（CPS）, typical structure of cyber-physical-energy systems(CPES), information perception of CPES, optimal schedule of power system with renewable generation, control and optimization of enterprise energy system and building energy system, and CPS comprehensive security.

7.9.2　课程基本情况（Basic Information of the Course）

课程名称	信息物理融合能源系统 Cyber-Physical Energy Systems							
开课时间	一年级		二年级		三年级		四年级	
	秋	春	秋	春	秋	春	秋	春

开课时间（续）	总学分	2	总学时	32

课程定位	储能科学与工程专业本科生储能系统课程群选修课
授课学时分配	课堂讲授32学时
先修课程	运筹学,自动控制理论,储能系统检测与估计
后续课程	无
教学方式	课堂教学,实验与综述报告,讨论
考核方式	大报告占70%,平时成绩占30%
参考教材	[1] 管晓宏,赵千川,贾庆山,等.信息物理融合能源系统.北京:科学出版社,2016.
参考资料	[1] 钱学森,宋健.工程控制论.新世纪版.北京:科学出版社,2011. [2] 阿卢尔.信息物理融合系统（CPS）原理.董云卫,译.北京:机械工业出版社,2017.

7.9.3　教学目的和基本要求（Teaching Objectives and Basic Requirements）

（1）使学生具备扎实的数学基础，培养学生运用系统工程思想方法分析和解决实际能源系统的意识和能力，使学生掌握信息感知、能源电力系统优化、信息物理融合能源系统综合安全等信息物理融合能源系统的基础理论、概念和相关算法。

（2）通过接受控制科学与工程、电气工程、网络空间安全等多学科的研究方法训练，了解不同学科的特点及内涵，培养学生具有综合运用多学科基础理论、技术、研究方法的综合思维能力。

（3）培养学生能够使用计算机模拟来解决实际问题的能力，学会应用常见优化软件实现对实际问题的建模和求解；对信息物理融合能源系统领域的复杂科学与工程问题，能够进行分析、预测、模拟、求解和论证。

（1）Students are expected to improve their mathematical foundation and their awareness and ability to analyze and solve the actual energy system by using the system engineering method. Students can master the basic theories, concepts, and related algorithms of CPES, such as information perception, energy, and power system optimization, CPS, and comprehensive security of the energy system.

（2）Through the training of multi-disciplinary research methods such as control science and engineering, electrical engineering, cyberspace security, etc., students can understand the characteristics and connotations of different disciplines. This course also cultivates students' comprehensive thinking ability and skill of comprehensive application of multi-disciplinary basic theory, technology, and research methods.

（3）Students should have the ability to solve practical problems by using computer simulation. Students should learn to use common optimization software to realize modeling and solution of practical application problems, and analyze, predict, simulate, solve, and demonstrate complex scientific and engineering problems in the field of CPES.

7.9.4　课程大纲和知识点（Syllabus and Key Points）

第一章　绪论（introduction）

章节序号 chapter number	章节名称 chapters	知识点 key points	课时 class hour
1.1	信息物理融合系统 cyber-physical systems	（1）CPS与第四次工业革命 （2）CPS基本概念与内涵 （1）CPS and Industry 4.0 （2）basic concept and connotation of CPS	1

续表

章节序号 chapter number	章节名称 chapters	知识点 key points	课时 class hour
1.2	信息物理融合能源系统 cyber-physical energy systems	（1）信息物理融合能源系统的背景、意义和挑战 （2）能源系统工程 （1）background, significance, and challenge of CPES （2）energy systems engineering	1

第二章　CPES 的典型结构（typical structure of CPES）

章节序号 chapter number	章节名称 chapters	知识点 key points	课时 class hour
2.1	智能电网 smart grid	（1）智能电网概述 （2）智能电网的结构特点 （1）introduction to smart grid （2）structural characteristics of smart grid	1
2.2	能源互联网 energy internet	（1）能源系统节能优化问题及其挑战 （2）能源互联网的结构特点 （1）problems and challenges of energy-saving optimization of energy systems （2）structural characteristics of energy internet	1
2.3	能源系统源-网-荷-储的系统化结构 energy system's structure: source, transmission, storage, and demand	（1）系统化结构 （2）系统化仿真 （1）systematic structure （2）systematic simulation	2

第三章　CPES 的信息感知（CPES sensing）

章节序号 chapter number	章节名称 chapters	知识点 key points	课时 class hour
3.1	CPES信息感知框架 framework of CPES sensing	（1）CPES信息感知 （2）CPES信息感知方法 （1）CPES sensing （2）CPES sensing technology	2
3.2	典型的信息感知系统 typical sensing systems	（1）工业互联网 （2）典型数据通信网络 （1）industry internet （2）typical data communication network	3

第四章　含新能源的电力系统发电优化调度（optimal schedule of power system with renewable generation）

章节序号 chapter number	章节名称 chapters	知识点 key points	课时 class hour
4.1	风-储系统的运行特征 operating characteristics of wind-energy-storage systems	（1）风资源概述 （2）风-储系统运行结构 （1）overview of wind energy （2）operating structure of wind-energy-storage system	2
4.2	光-储系统的运行特征 operating characteristics of PV-storage systems	（1）光资源概述 （2）光-储系统运行结构 （1）overview of PV station （2）operating structure of PV-storage system	2
4.3	电力系统优化调度模型和优化方法 model and optimization method of power system	（1）电力系统优化调度模型 （2）电力系统优化调度方法 （1）schedule model of power system （2）schedule optimization method of power system	2

第五章　企业能源系统的控制与优化（control and optimization of enterprise energy system）

章节序号 chapter number	章节名称 chapters	知识点 key points	课时 class hour
5.1	企业能源系统 enterprise energy system	（1）高耗能企业概述 （2）高耗能企业储能单元 （3）企业EMS与MES （1）overview of energy-intensive enterprise（EIE） （2）storages in EIE （3）EMS and MES	1
5.2	企业能耗预测 energy consumption forecast	（1）企业能耗的基本特征 （2）企业能耗预测 （1）basic characteristics of enterprise energy consumption （2）enterprise energy consumption forecast	2
5.3	企业能源系统优化调度 energy schedule in EIE	（1）企业能源系统优化调度模型 （2）企业能源系统优化调度算法 （1）schedule model of enterprise energy system （2）schedule optimization method of enterprise energy system	2

第六章　楼宇能源系统的控制与优化（control and optimization of building energy system）

章节序号 chapter number	章节名称 chapters	知识点 key points	课时 class hour
6.1	楼宇能源系统 building energy system	（1）楼宇能源系统概述 （2）楼宇能源转换单元 （3）楼宇储能单元 （1）overview of building energy system （2）building energy conversion unit （3）building energy storage unit	2
6.2	楼宇能耗与人员舒适性 building energy consumption and personnel comfort	（1）人员舒适性评估 （2）楼宇能耗与人员舒适性的关系 （1）personnel comfort assessment （2）relationship between building energy consumption and personnel comfort	2
6.3	楼宇能源系统优化调度 schedule optimization of building energy system	（1）楼宇能源系统优化调度模型 （2）楼宇能源系统优化调度算法 （1）schedule model of building energy system （2）schedule optimization method of building energy system	2

第七章　CPES 综合安全（CPES integrated security）

章节序号 chapter number	章节名称 chapters	知识点 key points	课时 class hour
7.1	CPES综合安全的定义 definition of CPES integrated security	（1）CPES综合安全的典型案例 （2）CPES综合安全的定义 （1）cases on CPES integrated security （2）definition of CPES integrated security	2
7.2	CPES数据安全与纵深 防御 CPES data security and integrated defense	（1）CPES数据安全 （2）CPES纵深防御 （1）CPES data security （2）CPES integrated defense	2

_ 第 8 章 _

前沿讲座课程群

8.1 "大型储能工程导论"教学大纲

课程名称：大型储能工程导论

Course Title：Introduction to the Large-Scale Energy Storage Engineering

先修课程：储能原理 I，储能原理 II

Prerequisites：Energy Storage Principle I, Energy Storage Principle II

学分：1

Credits：1

8.1.1 课程目的和基本内容（Course Objectives and Basic Contents）

本课程聚焦典型大型储能技术的研究前沿与工程案例，拓宽学生视野，培养学生分析问题和解决问题的能力，为后续深入学习储能领域相关知识以及毕业后从事相关科研工作奠定基础。大型储能工程导论主要内容包括大型储能技术简介、热质储能技术前沿、电化学与电磁储能技术前沿、储能系统技术前沿等。本课程紧跟储能技术的发展前沿，不断更新储能技术前沿内容。

The course focuses on the research frontier and typical engineering cases of large-scale energy storage technologies, thereby extends students' thinking, develops students' ability to analyze and solve problems, and allows students to lay a solid foundation for future study and research in energy storage. Introduction to the Large-Scale Energy Storage Engineering includes an overview of large-scale energy storage technologies, frontiers in heat and mass energy storage technologies, frontiers in electrochemical and electromagnetic energy storage technologies, and frontiers in energy storage systems. This course pursues the development of large-scale energy storage technologies and updates the frontier contents of large-scale energy storage technologies.

8.1.2　课程基本情况（Basic Information of the Course）

课程名称	大型储能工程导论 Introduction to the Large-Scale Energy Storage Engineering							
开课时间	一年级		二年级		三年级	四年级	总学分	1	总学时	16

课程名称	大型储能工程导论 Introduction to the Large-Scale Energy Storage Engineering
开课时间	一年级　二年级　**三年级**　四年级　　总学分 1　总学时 16 秋　春　秋　春　**秋　春**　秋　春
课程定位	储能科学与工程专业本科生前沿讲座课程群必修课
授课学时分配	课堂讲授16学时
先修课程	储能原理I，储能原理II
后续课程	热质储能综合实验，储能电池设计、制作及集成化实验，储能装置设计与开发实验
教学方式	课堂讲授，文献阅读与小组讨论，大作业
考核方式	大作业成绩占80%，平时成绩占20%
参考教材	[1] 丁玉龙，来小康，陈海生. 储能技术及应用. 北京：化学工业出版社，2018.
参考资料	[1] 黄志高，储能原理与技术. 北京：中国水利水电出版社，2018. [2] 饶中浩，汪双凤. 储能技术概论. 北京：中国矿业大学出版社，2017.

8.1.3　教学目的和基本要求（Teaching Objectives and Basic Requirements）

掌握储能技术的分类和应用范围，熟悉储能技术的现状和发展趋势，熟悉大规模储能工程的典型案例。

Master the types and applications of energy storage technologies. Be familiar with the application status and development trends of energy storage technologies. Be familiar with the typical application cases of large-scale energy storage technologies.

8.1.4　课程大纲和知识点（Syllabus and Key Points）

第一章　大型储能技术简介（introduction to large-scale energy storage technologies）

序号 number	标题 titles	提纲 outlines	课时 class hour
1.1	为什么需要发展大型储能技术 reasons to develop large-scale energy storage technologies	（1）高效利用可再生能源的需要 （2）电力系统高效安全运行的需要 （3）高效利用工业余热的需要 （1）for efficient utilization of renewable energy （2）for efficient and safe operation of grids （3）for efficient utilization of industrial waste heat	0.5

续表

序号 number	标题 titles	提纲 outlines	课时 class hour
1.2	储能技术的种类、学科背景及应用范围 classifications of energy storage technologies, their academic disciplines and application ranges	（1）储能技术的分类 （2）不同储能技术的学科背景 （3）不同储能技术的应用范围 （1）types of energy storage technologies （2）academic disciplines of different energy storage technologies （3）application range of different energy storage technologies	0.5
1.3	目前国内外大规模储能技术应用概况 applications overview of domestic and international large-scale energy storage technologies	（1）我国储能技术应用实例 （2）国外储能技术应用实例 （1）domestic application cases of large-scale energy storage technologies （2）oversea application cases of large-scale energy storage technologies	0.5
1.4	大规模储能工程的研究方向 research trend in large-scale energy storage technologies	根据报告时的情况而定 to be determined when the lecture is going to be provided	0.5

第二章 热质储能技术前沿（frontiers in heat and mass energy storage technologies）

序号 number	标题 titles	提纲 outlines	课时 class hour
2.1	储能型太阳能光热发电系统与应用 systems and applications of energy-storage-based solar thermal power generation technology	（1）储热技术简介 （2）储能型太阳能光热发电系统介绍 （3）储能型太阳能光热发电系统的发展 （4）储能型太阳能光热发电系统典型应用实例 （1）overview of heat storage technologies （2）introduction of energy-storage-based solar thermal power generation systems （3）development of energy-storage-based solar thermal power generation systems （4）typical application cases of energy-storage-based solar thermal power generation systems	2
2.2	压缩空气储能系统与应用	（1）压缩空气储能系统介绍 （2）压缩空气储能系统的发展 （3）压缩空气储能系统典型应用实例	2

序号 number	标题 titles	提纲 outlines	课时 class hour
2.2	energy storage systems and applications of compressed air technology	（1）introduction of compressed air energy storage systems （2）development of compressed air energy storage systems （3）typical application cases of compressed air energy storage systems	2
2.3	氢燃料电池系统与应用 systems and applications of hydrogen fuel cells	（1）氢能与储氢技术简介 （2）氢燃料电池介绍 （3）氢燃料电池的发展 （4）氢燃料电池典型应用实例 （1）overview of hydrogen energy and hydrogen storage technologies （2）introduction of hydrogen fuel cells （3）development of hydrogen fuel cells （4）typical application cases of hydrogen fuel cells	2

第三章　电化学与电磁储能技术前沿（frontiers in electrochemical and electromagnetic energy storage technologies）

序号 number	标题 titles	提纲 outlines	课时 class hour
3.1	超级电容器储能系统与应用（二选一） energy storage systems and applications of supercapacitors(alternative)	（1）超级电容器储能系统介绍 （2）超级电容器储能系统的发展 （3）超级电容器储能系统典型应用实例 （1）introduction of supercapacitor energy storage systems （2）development of supercapacitor energy storage systems （3）typical application cases of supercapacitor energy storage systems	2
3.2	锂离子电池储能系统与应用（二选一） energy storage systems and applications of lithium-ion batteries（alternative）	（1）锂离子电池介绍 （2）锂离子电池储能系统的发展 （3）锂离子电池储能系统典型应用实例 （1）introduction of lithium-ion battery energy storage systems （2）development of lithium-ion battery energy storage systems （3）typical application cases of lithium-ion battery energy storage systems	2

续表

序号 number	标题 titles	提纲 outlines	课时 class hour
3.3	超导储能系统与应用 systems and applications of superconducting magnetic energy storage	（1）超导储能系统介绍 （2）超导储能系统的发展 （3）超导储能系统典型应用实例 （1）introduction of superconducting magnetic energy storage systems （2）development of superconducting magnetic energy storage systems （3）typical application cases of superconducting magnetic energy storage systems	2

第四章 系统储能技术前沿（frontiers in energy system storage technologies）

序号 number	标题 titles	提纲 outlines	课时 class hour
4.1	电力储能装备的发展 development of power energy storage equipments	（1）电力储能装备的作用 （2）电力储能装备的类型与基本结构 （3）电力储能装备的发展 （4）电力储能装备典型应用案例与分析 （1）functions of power energy storage equipments （2）types and basic structures of power energy storage equipments （3）development of power energy storage equipments （4）typical application cases and analysis of power energy storage equipments	2
4.2	储能在电力系统的典型应用 applications of energy storage in power system	（1）储能的典型运行模式 （2）储能在发电侧典型应用案例与分析 （3）储能在输配侧典型应用案例与分析 （4）储能在用户侧典型应用案例与分析 （5）储能在电力市场典型应用案例与分析 （1）typical operation models of energy storage （2）typical application cases and analysis of energy storage in power generation system （3）typical application cases and analysis of energy storage in transmission and distribution system （4）typical application cases and analysis of energy storage in demand-side （5）typical application cases and analysis of energy storage in electricity market	2

_ 第 9 章 _

专业综合实验课程群

9.1 "热质储能综合实验"教学大纲

课程名称：热质储能综合实验

Course Title：Comprehensive Experiment of Thermal-Mass Energy Storage

先修课程：储能原理 II，储能热流基础（含实验），传热传质学，先进热力系统技术及仿真

Prerequisites：Energy Storage Principle II, Fundamentals of Thermal-Fluid Science in Energy Storage（with Experiments）, Heat and Mass Transfer, Principle and Simulation of Advanced Thermodynamic Systems

学分：1

Credits：1

9.1.1 课程目的和基本内容（Course Objectives and Basic Contents）

本课程是面向储能科学与工程专业（热质储能模块）开设的一门集中实践课程。

课程以源 – 网 – 荷 – 储多能流离 / 并网仿真实践平台为依托，其任务是培养学生掌握能量的产生与转化、传输、储存与利用的基本规律，理解不同品位能源的不同利用形式。通过现场观摩及亲手操作，建立对储热 / 蓄冷、热功转换系统、液流电池系统、综合能源利用系统的直观认识；通过实验测试，掌握上述热质储能系统的基本运行规律及评价方法。

This course is a practical course for the undergraduates majoring in energy storage science and engineering (heat and mass energy storage module).

The course is based on the platform that physically simulates the multi-energy flow processes between source-charge-storage modes regarding off-grid and on-grid conditions. It trains students to master the basic laws of energy generation, conversion, transfer, storage and usage, and understand the different utilization forms of various grades of energy. Through observation and hands-on operation, the course establishes an intuitive understanding of important concepts for students, including heat storage/cold storage, heat to power conversion system, flow battery system, and integrated energy utilization system. Through experimental operation, the students are expected to master the basic operating rules and evaluation methods of the above-mentioned heat and mass energy storage systems.

9.1.2 课程基本情况（Basic Information of the Course）

课程名称	热质储能综合实验 Comprehensive Experiment of Thermal-mass Energy Storage									
开课时间	一年级		二年级		三年级		四年级		总学分	1	总学时	32
	秋	春	秋	春	秋	春	秋	春				
课程定位	储能科学与工程专业本科生专业综合实验课程群（热质储能模块）必修课									
授课学时分配	实验32学时									
先修课程	储能原理Ⅱ,储能热流基础（含实验）,传热传质学,先进热力系统技术及仿真									
后续课程	无									
教学方式	实验									
考核方式	实验总结报告占60%,合作、创新等占40%									
参考教材	无									
参考资料	无									

9.1.3 实验目的和基本要求（Experiment Objectives and Basic Requirements）

通过本课程的学习,使学生对前期所学习的能量传递、转化与储存基本规律加深理解。
（1）掌握储热/蓄冷子系统在不同工况下的运行规律;
（2）掌握热功转换子系统在不同工况下的运行规律;
（3）掌握液流电池子系统在不同工况下的运行规律;
（4）掌握综合能源利用系统的概念,了解综合能源利用系统基本运行规律及评价方法。

Through the study of this course, students will have a deeper understanding of the basic laws of energy transmission, conversion and storage learned in the prerequisite courses.

（1）Master the operation characteristics of heat storage/cold storage subsystem under different working conditions.

（2）Master the operation characteristics of the heat to power conversion subsystem under different working conditions.

（3）Master the operating characteristics of the flow battery subsystem under different working conditions.

（4）Master the concept of the integrated energy utilization system, the basic operation rules, and evaluation methods of the integrated energy utilization system.

9.1.4 实验环节（Experiments）

序号 number	实验内容 experiment content	知识点 key points	课时 class hour
1	源-网-荷-储多能流离/并网仿真实践平台介绍 introduction of the physical simulation platform of multi-energy flow between source-charge-storage operated at off/on grid	（1）平台的组成及子系统 （2）操作规范 （1）composition and subsystems of the platform （2）operation standard	4
2	储热/蓄冷子系统实验研究 experimental research on heat storage/cold storage subsystem	（1）储热/蓄冷子系统介绍 （2）储热/蓄冷子系统性能评价方式及其计算方法 （3）储热/蓄冷子系统实验测试 （4）储热/蓄冷子系统经济性评价 （1）introduction of heat storage/cold storage subsystem （2）performance evaluation method and calculation method of heat storage/cold storage subsystem （3）experimental test of heat storage/cold storage subsystem （4）economic evaluation of heat storage/cold storage subsystem	6
3	热功转换子系统实验研究 experimental research on heat to power conversion subsystem	（1）热功转换子系统介绍 （2）热功转换子系统性能评价方式及其计算方法 （3）热功转换子系统实验测试 （4）热功转换子系统经济性评价 （1）introduction of heat to power conversion subsystem （2）performance evaluation method and calculation method of heat to power conversion subsystem （3）experimental test of heat to power conversion subsystem （4）economic evaluation of heat to power conversion subsystem	6

续表

序号 number	实验内容 experiment content	知识点 key points	课时 class hour
4	液流电池子系统实验研究 experimental research on flow battery subsystem	（1）液流电池子系统介绍 （2）液流电池子系统性能评价方式及其计算方法 （3）液流电池子系统实验测试 （4）液流电池子系统经济性评价 （1）introduction of flow battery subsystem （2）performance evaluation method and calculation method of flow battery subsystem （3）experimental test of flow battery subsystem （4）economic evaluation of flow battery subsystem	6
5	综合能源利用系统实验研究 experimental research on integrated energy utilization system	（1）综合能源利用系统介绍 （2）综合能源利用系统性能评价方式及其计算方法 （3）综合能源利用系统实验测试 （4）综合能源利用系统经济性评价 （1）introduction of integrated energy utilization system （2）performance evaluation method and calculation method of integrated energy utilization system （3）experimental test of integrated energy utilization system （4）economic evaluation of integrated energy utilization system	10

9.2　"储能电池设计、制作及集成化实验"教学大纲

课程名称：储能电池设计、制作及集成化实验

Course Title：Design，Manufacture and Assembly of Batteries

先修课程：储能化学基础（含实验），电化学基础

Prerequisites：General Chemistry for Energy Storage，Fundamentals of Electrochemistry

学分：1

Credits：1

9.2.1　课程目的和基本内容（Course Objectives and Basic Contents）

本课程是储能科学与工程专业的一门专业综合性实验课。

化学储能电池中具有代表性的锂离子电池，由于工作电压高、体积小、质量小、比能量高、无记忆效应、无污染、自放电小、循环寿命长等优点，近年来在一些轻、薄、小的

电子设备上已得到了很好的应用，另外在动力能源上也有长足的发展。

　　本课程的主要内容是通过锂离子正极材料的制备、纽扣电池的封装及电池充放电性能的测量，全面理解和掌握锂离子电池的原理与制备工艺。

This course is a professional comprehensive experimental class for undergraduates majoring in energy storage science and engineering.

Lithium-ion batteries, which are representative of chemical energy storage batteries, have advantages such as high working voltage, small size, light weight, high specific energy, no memory effect, pollution-free, low self-discharge, and long cycle life. In recent years, it has been widely adopted in some light, thin, and small electronic devices, and there has also been considerable development in power systems.

The main content of this course is to comprehensively understand and master the principle and preparation process of lithium-ion batteries through the preparation of lithium-ion cathode materials, the packaging of button batteries, and the measurement of battery charge and discharge performance.

9.2.2 课程基本情况（Basic Information of the Course）

课程名称	储能电池设计、制作及集成化实验 Design、Manufacture and Assembly of Batteries										
开课时间	一年级		二年级		三年级		四年级		总学分	1	总学时	32
	秋	春	秋	春	秋	春	秋	春				
课程定位	储能科学与工程专业本科生专业综合实验课程群（电化学与电磁储能模块）必修课											
授课学时分配	实验32学时											
先修课程	储能化学基础（含实验），电化学基础											
后续课程	无											
教学方式	讨论,实验,报告,答辩											
考核方式	设计方案+方案答辩+方案实施+项目报告答辩+综合评价											
参考教材	[1] 郭炳焜,李新海,杨松青.化学电源:电池原理及制造技术.长沙:中南大学出版社,2000. [2] 吴宇平,万春荣,姜长印,等.锂离子二次电池.北京:化学工业出版社,2002.											
参考资料	无											

9.2.3 实验目的和基本要求（Experiment Objectives and Basic Requirements）

实验目的：

通过锂离子正极材料的制备、扣式电池的封装及电池充放电性能的测量，全面理解和掌握锂离子电池的原理与制备工艺。

基本要求：

（1）掌握锂离子电池的结构及充放电原理；

（2）掌握锂离子正极材料的制备过程及工艺；

（3）掌握锂离子电池的封装工艺；

（4）了解正极材料改性的目的、意义及方法；

（5）掌握掺杂离子改性的原理；

（6）掌握锂离子电池的充放电过程及测试方法。

Experiment Objectives：

The main purpose of this course is to fully understand and master the principle and preparation process of lithium-ion batteries through the preparation of lithium-ion cathode materials, the packaging of button cells, and the measurement of battery charge and discharge performance.

Basic Requirements:

（1）Master the structure and charging and discharging principle of lithium-ion batteries.

（2）Master the preparation process of lithium-ion cathode materials.

（3）Master the packaging process of button batteries.

（4）Understand the purpose, significance, and methods of modification of cathode materials.

（5）Master the principle of doping ion modification.

（6）Master charging and discharging process and test methods of lithium-ion batteries.

9.2.4　实验环节（Experiments）

序号 number	实验内容 experiment content	知识点 key points	课时 class hour
1	锂离子正极材料的制备 preparation of lithium-ion cathode materials	（1）锂离子电池的结构 （2）锂离子电池的充放电原理 （3）锂离子正极材料的制备过程及工艺 （1）structure of lithium-ion batteries （2）charging and discharging principles of lithium-ion batteries （3）preparation process and technology of lithium-ion cathode materials	6
2	扣式电池的封装 packaging of button cells	锂离子电池的封装工艺 packaging process of lithium-ion batteries	6

续表

序号 number	实验内容 experiment content	知识点 key points	课时 class hour
3	正极材料改性 cathode material modification	正极材料改性的目的、意义及方法 purpose, significance, and method to modify cathode materials	6
4	掺杂离子改性 doped ion modification	掺杂离子改性的原理 principle of doping ion modification	6
5	锂离子电池的充放电过程及测试方法 charging and discharging process and test method of lithium-ion batteries	锂离子电池的充放电过程及测试方法 charging and discharging process and test methods of lithium-ion batteries	8

9.3 "储能装置设计与开发实验"教学大纲

课程名称：储能装置设计与开发实验

Course Title：Energy Storage Device Design and Development Experiment

先修课程：电路，现代电子技术，储能系统设计

Prerequisites：Electric Circuit，Modern Electronic Technology, Design of Energy Storage System

学分：1

Credits：1

9.3.1 课程目的和基本内容（Course Objectives and Basic Contents）

本课程是面向储能科学与工程专业开设的一门集中实践课程。

课程以储能装置开发项目为背景，涉及锂离子电池、铅酸电池和超级电容储能装置的开发。培养学生掌握电池储能装置开发的基础知识和技能，可以根据设计系统的要求，独立完成锂离子电池、铅酸电池或超级电容储能装置的设计及开发，并给出经济性分析。

This course is a practical course for students majoring in energy storage science and engineering.

The course focuses on the development of energy storage devices based on lithium-ion batteries, lead-acid batteries, and supercapacitors. The objective of this course is to train students to master the fundamental knowledge and skills of battery energy storage device development. According to the requirements of system, students can independently develop an energy storage device based on lithium-ion battery, lead-acid battery, or supercapacitor. In addition, economic analysis for the device should be included in the final report.

9.3.2 课程基本情况（Basic Information of the Course）

课程名称	储能装置设计与开发实验 Energy Storage Device Design and Development Experiment							
开课时间	一年级	二年级	三年级	四年级	总学分	1	总学时	32
	秋　春	秋　春	秋　春	秋　春				
课程定位	储能科学与工程专业本科生专业综合实验课程群（系统储能模块）必修课							
授课学时分配	课堂讲授2学时+实验30学时							
先修课程	电路，现代电子技术，储能系统设计							
后续课程	无							
教学方式	课堂教学，实验与综述报告							
考核方式	实验总结报告占20%，项目完成情况占50%，汇报情况占20%，合作与创新等占10%							
参考教材	[1] 瑞恩，王朝阳.电池建模与电池管理系统设计.北京：机械工业出版社，2018.							
参考资料	[1] 吴福保，杨波，叶季蕾.电力系统储能应用技术.北京：中国水利水电出版社，2014. [2] 冯博琴，吴宁.微型计算机原理电力电子技术.北京：清华大学出版社，2011. [3] 张勇.ARM原理与C程序设计.西安：西安电子科技大学出版社，2006.							

9.3.3 实验目的和基本要求（Experiment Objectives and Basic Requirements）

通过本课程的学习，使学生掌握多种典型储能电池的储能装置的设计开发方法，为后续的研究打下良好的基础。

基本要求：

（1）熟悉锂离子电池、铅酸电池、超级电容储能装置的基本原理及技术参数；

（2）掌握锂离子电池、铅酸电池、超级电容器等单体电池的性能；

（3）分组完成基于锂离子电池（电压24 V、电流10 A）、铅酸电池（电压24 V、电流5 A）或超级电容器（电压24 V）储能装置的开发；

（4）完成相应储能装置的电池管理系统开发，并给出经济性分析。

Through the study of this course, students can master the design and development methods of energy storage devices based on various types of batteries, and lay a solid foundation for the further research.

Basic requirements:

（1）Being familiar with the basic principles and technical parameters of energy storage device based on lithium-ion batteries, lead-acid batteries, and supercapacitors.

（2）Mastering the cell performances of lithium-ion batteries, lead-acid batteries, and supercapacitors.

（3）Developing an energy storage device based on lithium-ion batteries (24 V, 10 A), lead-acid batteries (24 V, 5 A) or supercapacitors (24 V) in small group.

（4）The battery management system (BMS) should be developed，and the economic analysis should be included in the final report.

9.3.4　实验环节（Experiments）

序号 number	实验内容 experiment content	知识点 key points	课时 class hour
1	课程介绍 introduction to the project	（1）项目要求介绍 （2）实验装置介绍 （3）分组任选一种完成储能装置开发 （1）introduction to project requirements （2）introduction to the experimental equipment （3）completing the development of energy storage devices in any one of the groups	2
2	锂离子电池储能装置开发 development of lithium-ion battery energy storage device	（1）完成锂离子电池储能装置结构（电压24 V、电流10 A）的设计 （2）完成电池管理系统开发,含电池均衡系统、SoC估算 （3）对所开发装置给出经济性分析 （1）completing the design of lithium-ion battery energy storage device(24 V, 5 A) （2）developing a battery management system, including battery equilibrium system, estimation of SoC （3）completing economic analysis for the developed device	10
3	铅酸电池储能装置开发 development of lead-acid battery energy storage device	（1）完成铅酸电池储能装置结构（电压24 V、电流5 A）的设计 （2）完成电池管理系统开发,含电池均衡系统、SoC估算 （3）对所开发装置给出经济性分析 （1）completing the design of lead-acid battery energy storage device(24 V, 5 A) （2）developing a battery management system, including battery equilibrium system, estimation of SoC （3）completing economic analysis for the developed device	10

续表

序号 number	实验内容 experiment content	知识点 key points	课时 class hour
4	超级电容储能装置开发 development of supercapacitor energy storage device	（1）完成超级电容储能装置结构（电压24 V）的设计 （2）完成超级电容管理系统开发，含均衡系统、SoC估算 （3）对所开发装置给出经济性分析 （1）completing the design of supercapacitor energy storage device(24 V) （2）developing a supercapacitor management system, including equilibrium system, estimation of SoC （3）completing economic analysis for the developed device	10

2020 年注定是不平凡的一年。

对储能学科和产业来说，2020 年也是至关重要的一年，国家出台了一系列对储能学科和产业影响深远的政策。其中教育部、国家发展和改革委员会、国家能源局于 2020 年 1 月联合制定印发了《储能技术专业学科发展行动计划（2020—2024 年）》。我们在期刊《学科建设》上发表了《构建中国首个储能专业的探索与思考》，现亦刊印于此，以便于读者对我们创办储能科学与工程专业的思路历程有进一步的了解。抛砖引玉，欢迎大家讨论并提出宝贵意见，我们共同努力把我国储能科学与工程本科专业办好。

载自：何雅玲，丁书江，宋政湘，等. 构建中国首个储能专业的探索与思考. 学科建设. 2020，2（1）：34–39.

构建中国首个储能专业的探索与思考

何雅玲，丁书江，宋政湘，徐友龙，兰　剑

（西安交通大学，陕西，西安710049）

【摘　要】储能技术专业建设是应对全球能源格局变化、争取世界能源领域竞争主动权、推动我国储能产业和新能源领域高质量发展的现实需要和必然选择。本文阐述了构建储能专业的意义及所面临的挑战，聚焦西安交通大学建设中国高校首个储能专业方面的研究与探索，为我国高校建设储能技术相关专业提供参考案例。

【关键词】储能，专业建设，学科交叉

一、引言：世界能源格局的深刻变革和中国能源发展的新需求

当前，全球能源格局正在发生由依赖传统化石能源向追求清洁高效能源的深刻转变，新一轮能源革命蓬勃兴起，非常规油气、低碳能源、可再生能源、安全先进核能等一大批新兴能源技术正在改变传统能源格局。随着我国经济总量和人民生活水平的持续提升，我国的能源结构也正经历前所未有的深刻调整，我国能源资源发展将从总量扩大转向提质增效的新阶段，并将发挥更为重要的民生保障和经济推动作用。

近年来，对新能源和可再生能源的研究和开发，寻求提高能源利用率的先进方法，已成为全球共同关注的首要问题，也成为我国加快能源领域供给侧结构性改革的重要内容。中国作为一个能源生产和消费大国，既有节能减排的需求，也有能源增长以支撑经济发展

的需要，这就需要大力发展储能技术和储能产业。储能技术是新能源发展的核心支撑，是实现可再生能源大规模接入，提高电力系统效率、安全性和经济性的关键技术；储能产业作为新能源发展的核心支撑，关系到能源、交通、电力等多个重要行业的发展，尤其在当今能源枯竭日益加剧、能源消费供求不平衡的大环境下，储能可以突破传统能源模式时间与空间的限制，是能源结构转型的关键和推手，已成为主要发达国家竞相发展的战略性新兴产业。

近年来中国储能产业在项目规划、政策支持和产能布局等方面均加快了发展的脚步，未来几年随着可再生能源行业的快速发展，储能市场亦将迎来快速的增长。但是我国储能产业还处于发展的初级阶段，尚以示范应用为主，储能商业化应用面临着储能成本偏高、电力交易市场化程度不健全、储能技术路线不成熟、缺乏储能价格有效激励等各方面的问题，可谓机遇与挑战共存。

二、"储能科学与工程"专业建设的重要意义

在世界能源格局的深刻变动中，我国要保证能源安全，促进储能产业的关键核心技术攻关和自主创新能力，在储能产业的世界竞争中追赶超越、占据先机，需要以人才培养为立足点，以产教融合发展推动储能领域高精尖人才的培养。立足产业发展重大需求，统筹整合高等教育资源，加快培养急需紧缺人才，破解共性和瓶颈技术，建设独立的"储能科学与工程"专业是推动我国储能产业和能源高质量发展的必然选择。

2016 年 5 月，中共中央、国务院发布《国家创新驱动发展战略纲要》。该纲要指出，"发展安全清洁高效的现代能源技术，推动能源生产和消费革命。……加快核能、太阳能、风能、生物质能等清洁能源和新能源技术开发、装备研制及大规模应用，攻克大规模供需互动、储能和并网关键技术。"在响应国家的重大战略部署、推动能源产业深度优化革新的背景下，为加快培养储能领域"高精尖缺"人才，教育部、国家发展和改革委员会、国家能源局联合制定了《储能技术专业学科发展行动计划（2020—2024 年）》。该计划提出，"支持有关高校围绕产业需求、结合办学定位、整合办学优势，布局建设储能技术、储能材料、储能管理等新专业，改造升级材料物理、材料化学、新能源科学与工程、新能源材料与器件等已有专业。"希望经过 5 年左右努力，增设若干储能技术本科专业、二级学科和交叉学科，实现相关学科多领域交叉融合、协同创新，加快培养急需紧缺人才，破解共性和瓶颈技术。

我国高校在储能方向上的人才培养具有一定基础，一些学校如西安交通大学、浙江大学、华北电力大学等于 2011 年开设了新能源科学与工程专业，电子科技大学、中南大学和四川大学于 2011 年也开设了新能源材料与器件专业。但目前国内外高校均未将储能设立为单独专业，仅在能源相关各专业中进行相对独立的人才培养，难以突破学科专业壁垒形成合力，以支撑储能领域的发展。此外，传统工科教育的培养方式、方法存在偏向理论教学，实践性与场景性不够强，学生实践操作及创新能力不足的现象，较难满足新兴产业对于人才的需求。

2020 年 2 月 21 日，西安交通大学申请增设的全国首个且当年唯一一个储能科学与工

程专业获教育部批准。如何加快推进储能相关专业建设、完善储能科学与技术学科专业宏观布局、探索储能相关专业人才培养理念及培养模式，是国内拟设储能相关专业的高校需要思考的共同问题。本文结合储能科学与工程专业建设的特点，聚焦储能科学与工程专业建设的目的与问题，介绍西安交通大学在建设理念、知识体系构建、人才培养模式等方面的革新与实践。同时，结合新时代发展的规律与特点，构建产教融合、学科交叉、面向战略新兴产业与引领性行业的人才培养模式，为实现高校更高质量的发展，提供经验借鉴。

三、"储能科学与工程"专业建设的问题与挑战

储能科学与工程专业在申请阶段经过了审慎的调研与探讨，然而就现阶段而言，储能科学与工程专业在建设过程中还面临如下问题与挑战。

1. 如何突破传统工科模式，培养适应新产业需求的储能专业人才

储能专业研究虽然已有较长历史，但是国内外大学均未将储能设为单独专业，专业建设无先例可循。在无先进经验可借鉴的情况下，高校在新专业建立的过程中容易陷入传统培养模式思维。在设定培养目标时，对储能产业、企业人才需求甚至自身特色缺乏足够深入的了解，导致人才培养体系存在同企业、创新实践、经济社会需求脱节的问题。而且如果储能专业照搬传统工科的培养目标、课程体系和办学模式，将导致高校人才培养目标缺乏多样性和竞争力，与国家重大工程建设、企业技术创新需求相脱节。所以，如何重构核心知识，升级培养模式，将科学、人文、工程交叉融合，培养具备成为面向"全工程"、面向"复杂"应用领域甚至是人文科学的领袖人物，从而主动服务国家、社会、新产业人才需求，成为储能专业建立的关键问题。

2. 如何构建适应专业拓宽延伸、多学科交叉融合的储能专业人才培养体系

现有储能技术不成熟，产业链较长，单一储能技术的研发相对孤立。然而，储能技术涉及物理学、化学、材料、能源动力、电子信息、电气等学科领域知识，需要各学科高度融合。如何真正实现多学科交叉融合，形成完整储能专业培养体系，是创办储能专业亟待解决的问题。

3. 如何发展多方位优势资源，塑造校、企、研究院深度协同的育人模式

当前储能领域知识技术更新换代、成果转化速度加快，而高校教育仍相对滞后，课程知识体系不全且陈旧，加之教师重科研而实际工程背景不足，难以应对产业急速发展的新挑战以及新形势下企业对工程人才的需求。因此，人才培养还需突破壁垒，汇聚行业部门、高校、科研院所及企业等多方优势资源，不断完善科教融合、产教联动、校企合作的协同育人模式。如何拓展校、企、科研院所深度融合路径，深化多方协同合作层次，使各方资源使用效益最大化是建设储能专业需要面临的另一个重要问题。

四、"储能科学与工程"专业建设的基础

西安交通大学是我国最早兴办的著名高等学府，2017年学校入选国家"双一流"建设名单A类建设高校，动力工程及工程热物理、电气工程、机械工程、材料科学与工程、

力学、信息与通信工程、管理科学与工程、工商管理等 8 个学科入选一流建设学科。西安交通大学增设储能专业有着深厚的基础，具有增设储能学科、供给储能人才、推动区域能源行业技术转型、实现储能基础研究及技术应用突破的条件和能力。

在科学研究上，动力工程及工程热物理、电气工程、机械工程等大能源相关支撑学科具有全国领先的实力。其中动力工程及工程热物理、电气工程在第四轮学科评估中获 A+ 档评级，均位于全国前 2 名。相关支撑学科共拥有动力工程多相流国家重点实验室、电力设备电气绝缘国家重点实验室等 5 个国家重点实验室，流体机械国家专业实验室等 4 个国家专业或专项实验室，快速制造国家工程研究中心、流体机械及压缩机国家工程研究中心两个国家工程研究中心，以及海洋石油勘探国家工程实验室（联合）、高端制造装备协同创新中心等国家级平台。

在师资配备方面，主要牵头的能源与动力工程学院、电气工程学院、机械工程学院拥有中国科学院和中国工程院院士 8 名，国家级教学名师 4 名，"长江学者"特聘教授 29 名，"国家杰出青年科学基金"获得者 32 名，其他国家级人才计划 / 头衔 20 余名、国家级青年人才计划 30 余名，教育部人才 50 余名等。其他学院也有一批高层次人才直接或间接参与储能专业建设的过程当中。

在专业设置方面，现有的能源与动力工程专业、电气工程及其自动化专业在全国处领先地位，在国际上享有盛誉，其中能源与动力工程专业为国家综合改革试点专业、国家特色专业；新能源科学与工程专业是全国首批获批建立的战略性新兴产业相关本科专业，为国家特色专业；环境工程专业获批教育部"卓越工程师教育培养计划"。相关支撑学科拥有能源与动力工程专业国家级实验教学示范中心、核电厂与火电厂系统虚拟仿真教学实验中心等国家级实验教学中心为人才培养提供教学保障。

迁校以来，西安交通大学创造了百余项国内外能源科学研究领域的"第一"。如近三年学校获能源领域国家科技奖励 9 项，居全国高校第一，获批国家布局建设的唯一一个集能源基础前沿、关键技术、核心装备和能源战略研究为一体的综合性研究平台——国家西部能源研究院，获批能源领域目前唯一的"能源有序转化"国家自然科学基金基础科学中心等，实现了能源变革与可持续发展中的技术引领，为储能科学与工程专业建设提供了强有力的科研平台及基础支撑。

五、"储能科学与工程"专业建设新架构

西安交通大学自 2019 年筹建"储能科学与工程"专业以来，学校领导和专家多次赴国家发展和改革委员会、教育部汇报储能专业建设进展以及产业人才培养和创新方案推进情况，争取支持。与此同时，在主管副校长的领导下，教务处牵头，多次组织相关学院召开储能专业申报推进研讨会以及专业建设情况通报与工作布置会，确定了专业建设的新思路，开拓创新专业建设的新模式。

为做好储能专业建设，学校成立了以陶文铨院士为顾问，何雅玲院士为首席科学家，管晓红院士、别朝红教授等为委员的储能科学与工程专业建设专家委员会，并吸纳储能领域龙头企业包括国家电网公司、华电集团公司、华能集团公司、南方电网公司、宁德时代

新能源公司、比亚迪股份有限公司等的行业领军人物作为校外委员。委员会负责储能专业培养方案制订、课程体系建设、教材规划出版、教师团队构建、国家储能工程科教创新中心申办等各项专业建设相关的工作，构建课程教材、教学团队、实训平台、实践基地四位一体的创新建设体系，并针对储能科学与工程专业的特点设计建立革新的组织管理机制，旨在培养立足世界储能技术前沿、勇于创新的技术带头人以及具有宏观战略思维和市场思维的复合型管理人才，进而推动和加快国家及区域能源产业和经济发展。

1. 提出了先进的建设理念

学校面向国家能源战略，以储能行业迫切的人才需求为导向，以开放性、交叉性和创造性为原则，创新人才培养体制机制，整合学校六大储能相关学科的优势资源，联合国家储能领域十大龙头企业，深度合作、协同共建、全力打造世界一流的储能专业，为高等教育模式改革、新兴专业建设和储能技术的研究与开发提供新的范式。并在建设过程中坚持如下基本原则：

（1）坚持以国家战略和产业需求为导向

积极快速响应国家能源战略，针对储能行业迫切的人才需求，依托我校能源相关学科的雄厚实力，打造量身定做的人才培养方案。

（2）坚持多学科交叉的联合人才培养模式

打破学科壁垒，全面整合我校动力工程及工程热物理、电气工程、电子科学与技术、材料科学与工程、物理学、化学等六大储能相关学科（含两个 A+ 学科）的人才、平台、科研基地等资源，形成多学科联合共建储能专业的本、硕、博贯通培养新模式。

（3）坚持行业引领产教融合的专业建设机制

贯彻新工科理念，加强与储能行业十大企业的深度合作，共同构建培养方案，联合打造双师型队伍，合作建设实践实训平台与基地，促进高校人才培养和行业产业需求的有机衔接和深度融合，实现人才培养、科学研究和行业产业的共同发展。

2. 创新具有前瞻性的专业建设模式

储能科学与工程专业的建设瞄准国际先进水平，面向我国能源和产业发展对复合型储能人才需求，依托学校六个理工类优势学科，强强联合，打破专业壁垒、多学科交叉融合，以符合与时俱进的行业发展及创新型人才培养趋势为导向，设计了科学合理的目标方案，创新完备的专业体系，和切实可行的专业建设举措，初步打造了储能科学与工程专业建设模式，该建设模式突出了以下几个新特点。

（1）厚重数理基础，强调人文素养

储能科学与工程专业建设在传统能源专业基础上，以解决技术研发和生产制造过程中的复杂问题为导向，更加注重夯实数理基础理论知识。除此之外，还强调应培养储能人才良好的个人素养和从业操守、高度的社会责任感和国家使命感，要求储能人才能够将高效转化专业知识为国民经济做出贡献为己任，树立终生学习的理念，挖掘个人潜力，成长为领域的领军人才。

（2）强化理论联系实际，对接行业市场需求

学校广泛吸纳行业企业深度参与，共同完成培养方案和专业课程体系建设，联合国际化企业设立多学科交叉教学项目。一方面，聘请国际化企业的人员作为导师，深度参与人

才培养过程；另一方面，要求学生走出学校到企业中进行实践和锻炼，使其了解企业运行机制和企业文化，熟悉最新的工业技术发展和市场需求，从而加深学生对专业知识的理解并积累必要的社会经验，为其今后步入社会提前做好准备。

3. 创新面向攻克储能关键技术难题的专业建设体系

（1）科学的课程体系与教材建设

根据储能科学与工程专业的特点，针对储能技术存在的关键难题，在充分发挥学校相关学科与专业的优势的基础上，广泛吸纳行业企业深度参与人才培养工作，将专业分成热质储能、电化学和电磁储能、系统储能三个模块方向进行建设，并构建了通识教育、大类平台课、专业课和实践环节等有机结合的多层次课程体系，新增包括储能原理、储能材料基础、储能系统设计技术、储能系统安全管理、储氢技术及应用、储能系统及应用等多门核心课程，并新增储能电池设计技术、储能装置开发项目设计等在内的多门选修与实践课程。同时针对拟开设新课程进行相应的教材建设，形成完整的储能课程与教材体系，全面满足储能专业人才培养的需要。

（2）学科交叉融合创新性实训平台

储能科学与工程专业涉及多学科多领域交叉，为攻克储能专业关键技术难题，满足科研与教学协同培育创新人才的需求，打造多学科交叉融合的创新性实验实践平台——"国家储能科教创新中心"。实训平台以跨学院教学资源的共享为出发点，由学校统筹协调，实践教学中心统一创建储能专业所需的综合性和创新性实验实践平台，包括储能机理与材料创新研究实验平台、基础技术创新研究实验平台、综合应用创新研究实验平台和创新设计研究实验平台等，从而全方位地系统培养创新型的高层次拔尖人才。为满足平台建设的大物理空间的需求，学校在中国西部创新港为储能科学与工程专业特别预留了相应的物理空间，解决未来本专业高水平创新实验实践平台建设的问题。

（3）产教融合校外实践基地

学校拟与国家电网公司、南方电网公司、华能集团公司、华电集团公司、比亚迪股份有限公司等大型企业开展合作，建设10余个本专业的校外实践基地。推进学生到企业实习实训制度化、规范化，提高企业职工在岗教育培训覆盖水平和质量，并支持企业积极投入平台建设，探索开展订单式人才培养，不断创新储能技术产教融合实践基地建设管理机制，形成校企在人才培养方面稳定互惠的合作制度。

4. 创新校企共建的"双师型"团队

储能科学与工程专业以中国科学院院士、国家级教学名师何雅玲教授为专业负责人，现已吸纳学校能源与动力工程学院、电气工程学院、电子科学与工程学院、材料科学与工程、物理学院、化学学院等长期从事储能领域科研与教学的教师构成核心师资团队。依托学校现有的能源领域学科及师资队伍优势，未来将通过培养与引进方式，在5年内形成100人左右的高水平教学科研团队服务于专业建设。

在保证校内师资整体水平和体量的基础上，学校积极寻求与国家电网、中核集团等大型央企建立校企合作关系，促进校企人才双向交流，加强产教融合师资队伍建设。一方面落实教师定期到校外实践基地教学、岗位实践和学习制度，提高教师实践教学能力；另一方面探索实施产业教师"特岗计划"，从企业引进实践经验丰富、理论基础扎实的高级专业技术人员充实教师队伍，从而建立结构合理、专兼结合、科教产教融合的"双师型"高水平师资队伍。

5. 创新完备的质量监控与管理保障

研究制定多学科交叉融合背景下的学生能力评价标准和考核办法，建立质量监控体系。教学质量评价将采用教学观摩、座谈交流会、国内高校和企业同行评价以及学生评价相结合的方法，从教学日志、教学方法、教学内容、教学效果等方面多维度评价教师的教学质量。对学生同样采取多样化的考评方式，增加国际课程打分、国内外课程学分互认、科研小论文、分组讨论报告、实际工程问题分析、演讲、学科竞赛、企业实习效果等考评方式。同时借助学校大数据监控平台，对教学效果、教改成效和学生成绩等内容根据需求进行单独或综合数据分析。学校将从不同维度评价多学科培养模式的专业性和完整性，在前沿研究方向确立、研究方法融合、实验测试分析、技术开发和过程管理等方面深入分析研讨，最终制定一套可量化的、可反馈改进的、具有推广和应用价值的教学评估标准及实施细则，制定出较为合理的评价标准和规范。

集全校优势资源，探索和实践储能人才培养组织保障体系。由校企顶级专家构成的委员会负责专业内涵建设、培养方案论证、课程体系建设等工作。教务处牵头负责行政事务处理和多学科合作共建相关的协调工作。钱学森学院（西安交通大学拔尖创新人才培养试点学院）负责储能科学与工程专业的学生管理，并与专业相关学院共同完成教学计划的组织实施、学生的培养管理，以及任课教师的选聘等工作。实践教学中心负责实践实训平台的建设工作。

六、储能科学与工程专业建设的思考

1. 统筹规划储能科学与工程专业建设

专业建设应遵循一切从实际出发的原则，正确处理国家及地方经济发展对相关人力资源需求的关系，应充分发掘专业优势，适应区域经济发展规划与人才需求。与此同时，聚焦与学校办学定位和办学特色相匹配的学科方向，克服"功利性"和"碎片化"，专业建设方面要突出前瞻性、凸显储能科学与工程专业特色，走特色专业建设发展之路。

2. 强调多学科领域交叉

与传统能源科学相比，储能科学与工程专业更强调学科专业的交叉性和综合性。学科交叉是有选择的交叉渗透和综合集成，通过打破传统学科壁垒、打造高水平跨学科储能研究平台，从而凝聚自己的核心竞争力。不是学科专业的简单叠加，也不是无序杂乱的堆砌，而是把学校各个学科凸显优势特色的知识进行最优化的组装，凝练出新的储能发展方向。根据《储能技术专业学科发展行动计划（2020—2024年）》，一方面要加强能源动力工程及工程热物理、电气工程、电子科学与技术、物理学、化学等多学科内涵建设，促进储能技术与相关学科深度交叉融合；另一方面要吸纳各相关领域有影响力的专家教授参与对培养方案的反复修改与完善，如专业内课程设置、课程内容选择、学时学分安排等。依据建设理念不断改进，注意结合我国和本校教育的发展情况做出科学有效安排，尤其重视评价总结校与校、院与院、专业与专业间的交流和合作，优化培养方案、完善专业建设体系、探索变革人才培养模式。

3. 坚持科研教学融合联动

加强人才培养和科技创新的有机结合，建立健全以原始创新、集成创新和产业发展为导向的科技创新机制，以及大团队、大平台、大项目向学生深度开放的科研育人机制。整合学校多学科、多学院资源，包括国家级实验教学示范中心、教学基地等，打造多学科交叉融合的创新性实验实践平台"储能技术研究院"，满足科研与教学协同培育创新人才的需求，攻克关键技术难题。

4. 加强产业教学融合创新

整合高校、研究院、企业、行业资源，建设储能技术产业教学融合联合体，汇聚多方力量参与储能技术学科的专业建设。坚持共建、共享、共赢，打造创新储能技术产教融合实践基地，形成校企在人才培养方面稳定互惠的合作制度。推进"引企入校""引校进企"工作，吸纳行业企业深度参与人才培养工作，发挥企业在储能科学与技术高等教育人才培养中的重要主体作用，构建产教融合的创新生态系统，建立以企业为主体的协同创新和成果转化机制。

5. 扩大国际合作与交流

坚持学科建设国际化发展思路，不断丰富学科学术交流活动。通过发起、主办或协办国际会议，邀请国际顶尖专家来访讲学、开设精品大课堂，发起成立储能技术专业国际学术委员会，与国外高校和研究机构建立校级合作关系、开展联合培养项目、开设留学生班，增加赴国外访问、进修研究及攻读学位的教师和学生人次等方式不断扩大西安交通大学储能科学与工程专业的社会声誉和国际影响力。

参考文献

[1] 国家发展和改革委员会,国家能源局.关于印发《能源生产和消费革命战略（2016—2030）》的通知[EB/OL]. [2016–12–29].

[2] 教育部,国家发展和改革委员会,国家能源局.关于印发《储能技术专业学科发展行动计划（2020—2024年）》的通知［EB/OL].[2020–01–19].

[3] 丁明,陈忠,苏建徽,等.可再生能源发电中的电池储能系统综述[J]. 电力系统自动化,2013, 37(1): 19–25, 102.

[4] 刘英军,刘畅,玉伟,等.储能发展现状与趋势分析[J]. 中外能源,2017, 22(4): 80–88.

[5] 俞振华. 储能产业发展现状及趋势分析[A]. 第五届全国储能科学与技术大会摘要集. 2018.